The Medical Review Officer's Guide to Drug Testing

The Medical Review Officer's Guide to Drug Testing

Edited by

Robert B. Swotinsky, M.D., M.P.H.
Washington Occupational Health Associates, Inc.

JOHN WILEY & SONS, INC.
New York • Chichester • Weinheim • Brisbane • Singapore • Toronto

A NOTE TO THE READER:
This book has been electronically reproduced from digital information stored at John Wiley & Sons, Inc. We are pleased that the use of this new technology will enable us to keep works of enduring scholarly value in print as long as there is a reasonable demand for them. The content of this book is identical to previous printings.

Copyright © 1992, by John Wiley & Sons, Inc. All rights reserved.

Originally published as ISBN 0-442-00892-9.

Published simultaneously in Canada.

No part of this publication may be reproduced, stored in a retrieval system, or transmitted in any form or by any means, electronic, mechanical, photocopying, recording, scanning or otherwise, except as permitted under Sections 107 and 108 of the 1976 United States Copyright Act, without either the prior written permission of the Publisher, or authorization through payment of the appropriate per-copy fee to the Copyright Clearance Center, 222 Rosewood Drive, Danvers, MA 01923, (978) 750-8400, fax (978) 750-4744. Requests to the Publisher for permission should be addressed to the Permissions Department, John Wiley & Sons, Inc., 605 Third Avenue, New York, NY 10158-0012. (212) 850-6011, fax (212) 850-6008, E-mail: PERMREQ@WILEY.COM.

Library of Congress Cataloging-in-Publication Data

The Medical review officer's guide to drug testing /edited by Robert
  B. Swotinsky.
    p. cm.
  Includes bibliographical references and index.
  ISBN 0-471-28445-9
    1. Employees—Drug Testing. 2. Drug testing—Law and legislation—
  United States.   I. Swotinsky, Robert B.
  HF5549.5.D7M43 1992
  658.3'822—dc20   92-19472
                                                              CIP

*To my wife, Michelle,
who encouraged and supported me in this project.*

Contents

Foreword, xi
Preface, xvii
Acknowledgments, xix
Contributors, xxi

1 Drug Abuse in the Workplace   1
Joseph H. Autry III and Jamie B. Friedman

Federal Leadership Prior to 1900, 1
Prevalance of Drug Abuse in the Workplace, 5
Relationship of Drug Use to Performance
   and Productivity, 8
Components of a Comprehensive Drug-free
   Workplace Program, 10
Industry Responses to Drug Abuse in the
   Workplace, 13
Recent and Pending Legislation, 15
Sources for Additional Information, 20

2 Drug Testing in the Workplace   23
Robert Willette and Leo Kadehajian

History of Workplace Drug Testing, 23
Evaluating the Need for a Drug Testing Program, 31
Basic Elements of a Drug Testing Policy, 36
Evaluation of Drug Testing Programs, 46

3 Drug Testing Collection Procedures   53
Robert Swotinsky and Janet J. Beatey

Collection Site Preparation, 54
The Collection, 61
After Collection, 75
Breath and Blood Alcohol Testing, 78

4 Forensic Laboratory Drug Testing   81
Alan B. Jones and Michael A. Peat

Laboratory Selection, 81
Analytical Procedures, 83
The Drugs of Abuse, 87

viii Contents

    Alcohol, 93
    Hair Analysis, 93

5 The Medical Review Officer Function   97
Robert Swotinsky

    Standard Operating Procedure, 98
    Alcohol Testing, 128
    Rehabilitation and Return-to-Work
      Determinations, 128

6 Risk Management   133
William J. Judge and Robert Swotinsky

    The Legal Framework of Drug Testing:
      The Risk, 134
    Assisting Clients in Program Development, 135
    Risk Management Strategies, 136

7 Employee Assistance Programs   141
Larry V. Stockman

    Brief History, 141
    Description, 143
    EAP Education and Training, 145
    EAP Services Related to Drug Abuse, 145
    Guidelines for Referring Employees, 148
    Monitoring Recovery, 151
    Cost-Benefit Analyses, 151
    Selection Guidelines, 159

8 Monitoring Laboratory Performance   163
Dennis Crouch and Yale H. Caplan

    Internal Performance Monitoring, 166
    External Performance Monitoring, 171
    The Future of Monitoring Laboratory
      Performance, 188

9 Case Studies   193
Robert Swotinsky

    Programmatic Issues, 193
    Urine Collection Procedures, 195
    Pharmacologic Issues, 196
    The Medical Review Officer Function, 197
    Risk Management, 200
    Performance Testing, 201

*Appendix A: Drug Screening in the Workplace:
            Ethical Guidelines, 205*
*Appendix B: DHHS Mandatory Guidelines for
            Federal Workplace Drug Testing
            Programs, 207*
*Appendix C: DOT Procedures for Transportation
            Workplace Drug Testing Programs, 219*
Glossary, 231
Index, 239

Foreword

The field of occupational health witnessed sweeping developments during the 1980s, such as workplace hazard communication, ergonomics, financial incentives for healthy lifestyles, preferred provider arrangements, and managed care. Nowhere has change been more dramatic than in workplace drug testing. In 1980, drug testing was conducted by the military and the nuclear power industry, but by few other employers. Now, almost every Fortune 500 company prohibits illegal drug use on or off the job, and most require preemployment drug tests. An estimated 7 million Americans are subject to the federal government's drug testing requirements.

An entire industry has evolved in response to drug testing. Thousands of clinics and other providers collect specimens, hundreds of laboratories perform analyses, and thousands of physicians review and interpret the test results. Related services include employee assistance programs, policy development, and legal advice.

CONCERN FOR RELIABLE, CONFIDENTIAL TESTING PROCEDURES

As drug testing burgeoned, concerns arose about its legality and the potential invasion of individual privacy. Early legal challenges established the right of private employers to mandate drug testing. Several key U.S. Supreme Court cases established that public safety and security outweighed the rights of public employees in certain key jobs, paving the way for further expansion of drug testing.

The need to assure accurate and reliable drug testing procedures became paramount. Some early drug testing practices used relatively inaccurate analytic methods, or failed to use chain of custody collection procedures. Laboratory test results were usually sent directly to the employer without any medical interpretation. Professional organizations began setting standards. The American Society of Clinical Chemistry spearheaded a consensus that the "gold standard" of testing required both a screening test and a gas chromatography/mass spectrometry confirmatory test. In 1986, the American Occupational Medical Association issued ethical guidelines for drug testing that included a requirement for physician review of results, as follows:

> A qualified physician should evaluate positive results prior to a report being made to the employer. This may require the obtaining of supplemental information from the employee or applicant.

THE MEDICAL REVIEW OFFICER

The MRO role was first defined by the Department of Health and Human Services in its 1988 "Mandatory Guidelines for Federal Workplace Drug Testing Programs." These regulations require that all laboratory results be transmitted to a Medical Review Officer for review, who determines whether there are legitimate explanations for laboratory-positive results. (For example, a donor may have appropriately used a prescription medicine that could cause a positive drug test.) The MRO's determinations are then reported to the employer, who takes action according to internal company policy, federal requirements, or both.

Professional Society Support of the MRO Role

The MRO role has been strongly endorsed by professional medical societies, including the American College of Occupational and Environmental Medicine (ACOEM), the American Medical Association, and the American Society of Addiction Medicine. Policy statements have stated that the MRO should (1) be a licensed physician (MD or DO); (2) have at least a minimal training in subjects including chemical dependency illness (addiction), pharmacology of substance abuse, laboratory testing methodology and quality control, forensic toxicology, pertinent federal regulations, legal and ethical requirements, employee assistance programs and rehabilitation; and (3) maintain continuing education in this rapidly changing field. In May 1990, ACOEM commended the Department of Health and Human Services for insuring that licensed physicians evaluate the results of urine drug tests, and stated that "We believe that establishment of the Medical Review Officer (MRO) function alleviates fears among employees who are not using drugs that they will be falsely accused of being drug users."

Training of MROs

In 1989, the Department of Transportation (DOT) regulations took effect, mandating drug testing of millions of Americans. Suddenly there was a shortage of MROs and an economic opportunity for those who moved into this arena early. Prices for MRO services were generally high, and a number of companies providing nationwide MRO services quickly flourished.

In response to the need for qualified MROs, ACOEM developed a two-day MRO course in late 1989 and, with subsequent courses, had trained almost 2,000 MROs by December 1991. Two-day training courses sponsored by the ASAM, the Federal Aviation Administration, and other organizations have swelled the ranks of trained MROs to an estimated 5,000 physicians.

Credentialing of MROs

After focusing heavily on legal authority, confidentiality, laboratory quality control, chain of custody procedures, quality assurance, and other issues, policymakers have begun to focus on the quality of MRO services. Although there have been relatively few lawsuits pertaining to MRO competence, it is predicted that MROs will become more involved in litigation and, in turn, will be the focus of closer scrutiny.

Because of the complexity of federal regulations, and the host of gray areas that have been brought to light, federal officials have expressed growing concern about the training and credentialing of MROs. It would seem to make little sense to go to great effort and expense to achieve laboratory accuracy greater than 99.9 percent unless MROs can correctly review, interpret, and communicate these results to employers.

In 1992, two voluntary MRO certification programs emerged. The Medical Review Officer Certification Council (MROCC) was formed under the auspices of ACOEM. The American Association of Medical Review Officers (AAMRO) was sponsored by MRO Alert, publisher of a newsletter for MROs. Both organizations require completion of a formal MRO training course as an eligibility requirement and have developed a written certification examination. Efforts are under way to unify these organizations into a single MRO certifying body. Certifications limited to 3–5 years will encourage continuing education and periodic recertification.

THE NEED FOR EDUCATING PROVIDERS AND EMPLOYERS ON WORKPLACE DRUG TESTING

The need for better information about drug testing and the interpretation of results goes beyond physicians. Employers and other health professionals have expressed a strong desire to better understand the process and the complexity of medical decisionmaking involved in MRO work. *The Medical Review Officer's Guide to Drug Testing* offers the reader an understanding of:

- The nature and magnitude of the substance abuse problem in the United States and the workplace.
- An overview of drug testing in the workplace.
- A description of forensic urine drug test collection procedures.

- Forensic laboratory drug testing and techniques for monitoring laboratory performance.
- The MRO function as defined by federal requirements.
- Legal issues and risk management strategies.
- Elements of cost-effective employee assistance and rehabilitation programs.
- Clinical case studies relating common issues and their resolution.

FUTURE DIRECTIONS

Continued change in the world of drug testing is virtually certain. Alcohol has long been acknowledged as a more prevalent drug of abuse and a greater overall social burden than illegal drugs. In 1991, Congress mandated DOT to develop regulations governing alcohol testing in much the same fashion as testing for illegal drugs. With the anticipated arrival of final DOT regulations in late 1992, employers, laboratories, and MROs will enter a new era of alcohol testing. Because alcohol use is legal, interpretation of positive alcohol test results raises issues of fitness-for-duty rather than of legitimacy.

Proposed federal legislation would impose on all drug testing minimal federal standards that include chain of custody procedures, use of NIDA (National Institute on Drug Abuse) certified laboratories, and medical review of test results. On-site drug testing will gain increasing attention, especially for employees with safety-sensitive jobs such as airline pilots, train engineers, truck drivers, and nuclear power plant operators. Removing those who test positive to a screening test at the beginning of a work shift promises greater public safety than waiting several days for a urine screening and confirmation result.

Drug testing technology also continues to advance. Accurate detection is possible at lower cutoff levels than those currently in force. We can anticipate lower cutoff levels and the expansion of federally mandated drug testing to include more drugs of abuse.

PHILOSOPHICAL REFLECTIONS

Drug testing serves to deter drug abuse and to identify individuals at high risk of impairment in their work performance. Because of technological limitations, employers use the presence of drugs in the urine per se as a surrogate for determinations of impairment or the risk of future impairment. The physician serving as an MRO is, in this capacity, restricted from using clinical judgment about whether the individual was under the influence, impaired, addicted or an abuser of substances. The conscientious

MRO will inevitably be challenged by these limitations. While a better technology for evaluating these conditions would be a welcome replacement to drug testing, research efforts in areas such as neurobehavioral tests are in their infancy.

Drug abuse often reflects a dissatisfaction with one's life, be it personal or at work. Our society and employers must seek ways to foster healthier work environments—filled with creativity, challenge, and human concern ... in short, high-level wellness. Addressing these issues may help solve the problem of drug abuse, so that drug testing no longer remains a necessity.

KENT W. PETERSON, M.D., F.A.C.O.E.M.

Preface

The Medical Review Officer's Guide to Drug Testing is a comprehensive reference for health-care providers, employers, and others who wish to better understand this subject. This book emphasizes the practical, legal, and administrative aspects of workplace drug testing. It gives an overview of the drug problem in industry, reviews the federal rules regarding workplace drug testing (as of early 1992), and describes procedures by which drug test specimens are collected, analyzed, reviewed, and reported. The book also details the procedures by which laboratory performance is monitored, and describes the role of an employee assistance program in a drug-free workplace program.

This book is not intended as a critique on the wisdom of workplace drug testing. Drug testing is controversial, and its acceptability has been debated on legal, political, and technological grounds. Drug testing has, nevertheless, become widespread in recent years. In 1991, as many as 10 million drug tests were conducted in the United States. Recent legislation—the Omnibus Transportation Employee Testing Act of 1991—extends federal requirements for drug testing to millions of additional workers. This book addresses procedures that can help ensure that when drug testing is done, it is done correctly.

The technology, regulations, and case law of workplace drug testing have been rapidly evolving. This book was written in 1991, when the field had matured through several years of experience with the federal regulations. Subsequent developments will affect the practice of workplace drug testing. Therefore, this text should be considered an introduction to a subject that requires ongoing continuing education.

Acknowledgments

The contributors to this book deserve special thanks for sharing my enthusiasm for this project, and for their flexibility with respect to my editing for style, readability, and focus. I thank Ken Chase, who provided administrative support for this project, and from whom I learned my profession. I thank Kent Peterson, who coordinates the American College of Occupational and Environmental Medicine's MRO Training courses, which greatly influenced the design of this book. Finally, I thank my parents Jack and Anita Swotinsky for their long-term and dedicated support.

R.B.S.

Contributors

Joseph H. Autry III, M.D., is the Director, Division of Applied Research for the National Institute on Drug Abuse, within the Alcohol, Drug Abuse and Mental Health Administration (ADAMHA). He was formerly the Associate Administrator for Policy Coordination at ADAMHA, with responsibility for policy coordination and oversight, planning and evaluation, legislation, and the agency's AIDS programs. He has an extensive background in research and administration in mental health.

Janet J. Beatey, M.T. (A.S.C.P.), is Supervisor, Outpatient Centers, and Technical Assistant, Laboratory Operations at American Medical Laboratories, Inc. She coordinates collection services at multiple locations for a variety of drug testing programs.

Yale H. Caplan, Ph.D., is Clinical Professor and Director of Forensic Toxicology in the Department of Pathology of the University of Maryland School of Medicine. He also serves as Director of Forensic Toxicology and Director of Forensic Urine Drug Analysis for the National Center for Forensic Science, Division of Maryland Medical Laboratory, Inc. Dr. Caplan has served as a consultant to NIDA, the National Transportation Safety Board, and numerous other public and private organizations and companies.

Dennis Crouch, M.B.A., is Assistant Director of the Center for Human Toxicology and is a Research Instructor in the Department of Pharmacology and Toxicology at the University of Utah. He serves as an inspector for the laboratory certification programs of the College of American Pathologists and the National Institute on Drug Abuse, and was a member of the Monitoring Laboratory Performance working group at the 1989 NIDA Consensus Conference. During 1990 and 1991 he spent a sabbatical year at NIDA with the National Laboratory Certification Program.

Jamie B. Friedman, B.A., is a public health advisor with the Division of Applied Research at the National Institute on Drug Abuse, within the Alcohol, Drug Abuse and Mental Health Administration. She has a research background in drug abuse education and prevention.

Alan B. Jones, Ph.D., is Professor and Chair, Department of Pharmaceutics and Research Professor, Research Institute of Pharmaceutical Sciences, School of Pharmacy, University of Mississippi. He is active in drug abuse research and has served as an instructor in the training of inspectors for the National Laboratory Certification Program as well as an inspector in this program.

William J. Judge, J.D., LL.M., is counsel to the Chicago law firm of Feiwell, Galper & Lasky, Ltd. and general counsel to the nationwide consulting firm Workplace Information Network. He is also editor of the *Drug Testing Litigation Advisor* and research director for Policy Initiatives, Inc. He has written extensively on the legal issues of substance abuse and AIDS, has trained thousands of supervisors and employees nationwide, and is a frequent participant at seminars addressing these subjects.

Leo Kadehjian, Ph.D., is an independent consultant in Palo Alto, California, and has lectured and written on the pharmacology and testing of drugs of abuse. He has lectured worldwide, and has earned an oustanding speaker recognition from the American Association of Clinical Chemistry. He works closely with Duo Research, Inc., and Syva Company and has helped develop and maintain drug testing programs for major multinational corporations and for the U.S. courts.

Michael A. Peat, Ph.D., is Vice-President of Toxicology at CompuChem Laboratories, Inc., a large commercial laboratory dedicated to forensic-quality testing. He is a member of the Drug Testing Advisory Board to NIDA, and was Chairman of the Analytical Methods Working Group at NIDA's 1989 Consensus Conference. He also is an active contributor to the field of forensic toxicology through numerous publications, presentations, and leadership positions in professional organizations.

Larry V. Stockman, Ph.D., is International Accounts Manager for Health Affairs International, a provider of employee assistance services. He also is adjunct professor in the School of Finance & Business Administration, University of Houston at Clear Lake. Dr. Stockman has co-authored *Chemical Dependency: Recovery & Relapse Prevention* (in press) and *Adult Children Who Don't Grow Up* (Prima Press).

Robert B. Swotinsky, M.D., M.P.H., is a Senior Clinical Associate with Washington Occupational Health Associates, Inc., and is a Clinical Instructor of Medicine at the George Washington University School of Medicine. He coordinates medical surveillance and drug testing programs and serves as Medical Review Officer for multiple private- and public-sector clients on a nationwide basis. He has written and spoken extensively on the subject of workplace drug testing.

Robert Willette, Ph.D., is President of Duo Research, Inc., a consulting firm specializing in assisting private companies and government agencies in the development and implementation of drug control programs in the workplace, and the monitoring of drug testing procedures. He served as the Chief of the Research Technology Branch at the National Institute on Drug Abuse and continues to serve as a consultant to the U.S. Navy and other government agencies.

The Medical Review Officer's Guide to Drug Testing

1
Drug Abuse in the Workplace

Joseph H. Autry III, M.D., and
Jamie B. Friedman, B.A.

Many public and private employers have taken steps to address the problem of drugs in the workplace. Much of the motivation for implementing substance abuse programs in the workplace has come from federal initiatives and programs. At the same time, employers have been stimulated by the realities of life in late-twentieth-century America and find themselves in a global marketplace where this nation's rate of drug use is among the highest of any nation bringing goods and services into that marketplace. Such realities have prompted many private-sector employers to initiate drug-free workplace policies and programs prior to the development of federal guidelines and regulations.

In the 1960s, many private-sector employers developed policies and programs focused on alcohol abuse. However, as the problems of illicit drugs became more apparent, employers began focusing more on illicit drug use. In 1981, drug use among military personnel, and the serious consequences thereof, prompted the U.S. Department of Defense (DOD) to institute drug testing of all servicepeople. Drug testing became common in the civilian workplace in the 1980s, particularly among larger employers.

FEDERAL LEADERSHIP PRIOR TO 1990

Many private-sector initiatives were stimulated by federal government programs and policies that focus on illicit drug use. On September 15, 1986, President Reagan signed into law Executive Order 12564 (E.O. 12564),

---

The opinions expressed in this chapter are those of the authors and do not necessarily reflect the views of the National Institute on Drug Abuse.

which states, "The Federal government, as the largest employer in the nation, can and should show the way towards achieving drug-free workplaces through a program designed to offer drug users a helping hand."[1]

E.O. 12564 directed each federal agency to develop a comprehensive program to achieve a drug-free workplace for federal employees. Then President Reagan also presented it as a model for the private sector. It is a comprehensive model requiring that the prohibition against illegal drug use and its consequences be spelled out and effectively communicated to employees; that employees be educated about the dangers of drug use; that supervisors be trained concerning their responsibilities; that a helping hand in the form of an employee assistance program be available for employees who have a drug problem; and finally, that there be provisions for identifying illegal drug users, including drug testing on a controlled and carefully monitored basis.

The E.O is unequivocal in stating that federal employees must refrain from the use of illegal drugs; that their use, whether on or off duty, is contrary to the efficiency of the service; and that people who use illegal drugs are not suitable for federal employment. An agency may decide to provide a "safe harbor." That is, an agency may choose to take no action against employees who (1) voluntarily identify themselves as users, (2) obtain counseling or rehabilitation, and (3) thereafter refrain from using illegal drugs. However, in the absence of this set of conditions, agencies must initiate action to discipline any employee who is found to use illegal drugs. Further, agencies must initiate action to terminate any employee who uses illegal drugs and who refuses to stop or to get help.

E.O. 12564 and congressional Public Law 100-71[2] commissioned the Department of Health and Human Services (DHHS) to develop technical procedures that would make drug testing uniform, accurate, and confidential. The final DHHS Mandatory Guidelines for Federal Workplace Drug Testing Programs (also known as the "NIDA Guidelines," for the National Institute on Drug Abuse) were published in April 1988.[3] The Department of Transportation (DOT) adopted the NIDA Guidelines into its drug testing procedures, which were published in final form in December 1989.[4] The NIDA Guidelines and the DOT Procedures are recognized as the "gold standard" for reliable and legally defensible drug testing procedures.

In November 1988, the Drug-Free Workplace Act[5] was enacted by Congress. It conditioned that no federal agency is to enter into a contract or grant agreement with an individual or to enter into a contract in excess of $25,000 with entities other than individuals unless the contractor or grantee certifies that it will provide a drug-free workplace. It applies only to primary contractors and direct grantees and is implemented through the contracting and granting rules and procedures of each agency subject to Guidelines issued in May of 1990 by the Office of Management and Budget. The Drug-Free Workplace Act does not address drug testing, nor is it limited to workers in safety-sensitive positions. Under the Drug-Free Workplace Act, grantees and contractors must meet the following five conditions to achieve a drug-free workplace:

1. The contractor or grantee must publish a statement, a copy of which must be given to each employee, notifying him/her that unlawful manufacture, distribution, dispensation, possession, or use of a controlled substance is prohibited in the workplace and stating the actions that will be taken if that prohibition is violated.

2. The contractor or grantee must establish a drug-free awareness program of the dangers of drug abuse; the contractor's/grantee's policy on drug abuse; the availability of any employee assistance programs (EAPs), rehabilitation, and counseling services; and penalties for violation of the published prohibitions.

3. The contractor or grantee must specifically notify all employees in writing that, as a condition of employment, they must abide by the terms of the employer's published statement and notify the contractor or grantee of any conviction under a criminal drug statute for a violation occurring in the workplace no later than 5 days after the conviction. The employer must notify the relevant federal agency within 10 days after receiving notice.

4. The employer has 30 days after notification to take appropriate action, up to and including termination, or to require the employee to participate in an EAP or rehabilitation program.

5. The employer must continue good faith efforts to comply with the above requirements.

The penalty for making false certifications, for violation of certifications, or for failure to make good faith efforts to maintain a drug-free workplace are substantially the same for grants and contracts and may include suspension of payments, termination for default, and suspension or debarment for up to 5 years.

Also in 1988, the DOD issued regulations that require defense contractors to "institute and maintain a program for achieving the objective of a drug-free work force."[6] These regulations are aimed at ensuring national security through drug-free work environments, for example, those workplaces that affect the safety and reliability of weapons systems. The regulations apply to all contracts that require access to classified information, or where the contracting officer determines it necessary for reasons of national security, or to protect the health or safety of those who perform the contract or of those who use or are affected by the product of the contract. DOD regulations stipulate that contractors' drug-free workplace programs include: an EAP, supervisory training, provisions for self- and supervisory referrals to treatment, and provisions for identifying illegal drug users on a controlled and carefully monitored basis.

The contractors' employee drug testing program must test for the use of illegal drugs by employees in sensitive positions when there is reasonable cause: post-accident, return to duty following rehabilitation, and regular voluntary testing. DOD contractors are to test for those drugs prescribed by the NIDA Guidelines and must specify procedures for dealing with employees who test positive. In addition, the DOD regulations shall not apply to the extent that they are inconsistent with state or local law, or with an existing collective bargaining agreement. DOD enforces drug-free workplace requirements imposed on its contractors through normal remedies for noncompliance, including termination of the contract or suspension or debarment of the contractor.

On June 7, 1989, the Nuclear Regulatory Commission (NRC) published regulations affecting licensees authorized to operate nuclear power reactors.[7] Under these regulations, licensees are required to implement a Fitness for Duty Program that is aimed at creating a drug-free environment within nuclear power plants. The NRC regulations adopted most, but not all, of the provisions of the NIDA Guidelines with an awareness "that the Guidelines, as written by DHHS to apply to testing by federal agencies, do not perfectly fit the circumstances of the licensees regulated by the NRC."[7]

On December 1, 1989, the DOT published its final procedures[4] that impact approximately 4 million employees regulated by the six separate agencies within DOT: the Federal Railroad Administration, the Federal Highway Administration, the Federal Aviation Administration, the Research and Special Programs Administration (Natural Gas, Liquefied Natural Gas, and Hazardous Liquid Pipeline Operations), the Coast Guard, and the Urban Mass Transportation Administration (UMTA). Each of these agencies has issued separate rules applicable to the private-sector employers under its jurisdiction.* While there are differences in each agency's rule, they have the following provisions in common:

1. They require that collection and analysis of the specimen and review and reporting of the result be carried out according to the DOT Procedures.

---

*UMTA's antidrug program requirement was initially suspended by the U.S. Court of Appeals. In 1991, UMTA was given statutory authority to require drug testing.

2. They authorize testing for the same five drugs as the NIDA Guidelines.
3. They require employee education and supervisor training.
4. They require the availability of EAPs.
5. They permit employer discretion with regard to rehabilitation and discipline.
6. They require the following types of testing: reasonable cause, post-accident, periodic as part of scheduled medical examinations, preemployment for safety-sensitive positions, and random for safety-sensitive positions.
7. They require use of NIDA-certified laboratories.
8. They preempt state and local laws.

PREVALENCE OF DRUG USE IN THE WORKPLACE

The 1991 National Household Survey on Drug Abuse,[8] conducted by NIDA, showed a continuing decline in the use of most illicit drugs in the United States. The survey showed a 45 percent* reduction in current use of any illicit drug over the past six years and a 67 percent reduction in current cocaine use over the same period (current use is defined as having used an illicit drug at least once in the past month prior to the survey). However, the survey showed mixed trends: Current use of illegal drugs decreased overall. Current cocaine use increased by 18 percent over 1990, but remained 35 percent below the 1988 level.

There continue to be severe drug abuse problems in the United States. The 1991 National Household Survey estimated that 12.6 million Americans are current illicit drug users. The survey estimated that 855,000 people use cocaine at least once a week or more, virtually identical to the number found in the 1988 survey. Marijuana continues to be the most commonly used illicit drug, with an estimated 9.7 million Americans having used it during 1991.

Data from the 1991 National Household Survey show that 66 percent of those aged 18 and over reporting current illicit drug use are employed (see Figure 1-1), that is, approximately 7.5 million current drug users. Conversely, approximately 10 percent of full-time employees aged 18 to 34 report current use of illicit drugs. While this is an overall figure, there are significantly higher rates of use in certain demographic groups. In general, current drug use is more prevalent among male workers, younger workers, and workers with lower incomes. For example, data from the 1990 National Household Survey show that 17 percent of males aged 18-25 years (see Figure 1-2) and nearly 21 percent of males aged 18-34 years earning less than $9,000 (see Figure 1-3) are current users of illicit drugs.[9] Among 18- to 34-

---

*Percentages are rounded to the nearest whole number.

6 Drug Abuse in the Workplace

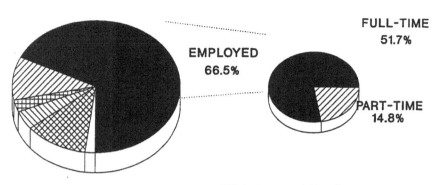

FIGURE 1-1. Current users of illicit drugs, aged 18 and over. *Source:* National Institute on Drug Abuse.[8]

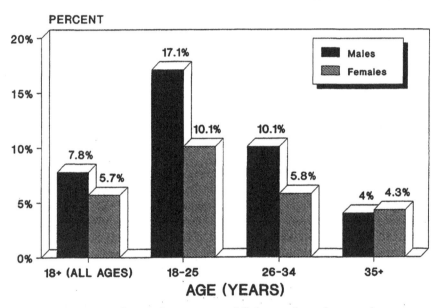

FIGURE 1-2. Current drug use among full-time employees by age and sex. *Source:* National Institute on Drug Abuse.[8]

year-old full-time employed Americans, 24 percent reported use of an illicit drug within the past year, and nearly 11 percent reported current use of an illicit drug. Of the full-time workers in this age range, 9 percent are current marijuana users and 2.1 percent are current cocaine users (see Figure 1–4).

The 1988 National Household Survey[10] found the lowest prevalence among retail workers (12%) and the highest prevalence among construction workers (22%) (see Figure 1-5). It should be noted that nearly 13 percent

Prevalence of Drug Use in the Workplace 7

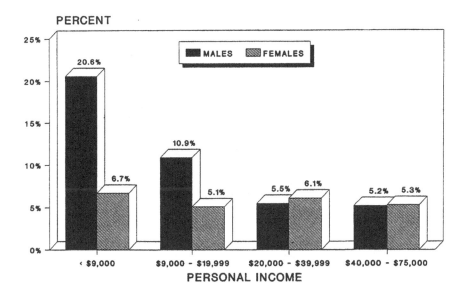

FIGURE 1-3. Current drug use among full-time employees 18–34 years old. *Source:* National Institute on Drug Abuse.[8]

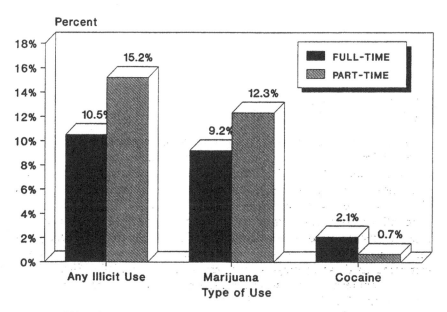

FIGURE 1-4. Current drug use among employees 18–34 years old. *Source:* National Institute on Drug Abuse.[8]

8   Drug Abuse in the Workplace

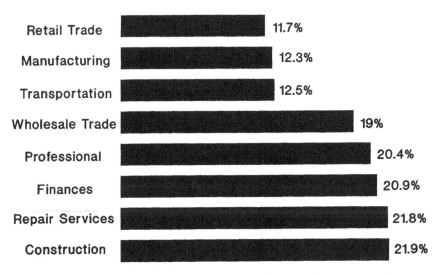

FIGURE 1-5.   Current drug use among full-time employees 18-34 years old, by industry. *Source:* National Institute on Drug Abuse.[9]

of the 4 million workers in the transportation industry report current use of illicit drugs. Also, 20 percent of individuals in the professional trades (most of whom do not fall under the federal regulations and encompass a broader range of individuals) report current use of illicit drugs.

RELATIONSHIP OF DRUG USE TO PERFORMANCE AND PRODUCTIVITY

As disturbing as these data are, only recently have we begun to study the potential impact of drug use on worker performance, productivity, and safety. In June 1991, the U.S. Postal Service conducted a follow-up study of employees testing positive for drugs at the time of preemployment urine drug testing versus those employees testing negative.[11] These employees were followed over 3.3 years of employment. During this time, absenteeism rates for those employees testing positive at the time of their preemployment urine drug test were 66 percent higher (11.39% versus 6.85%) when compared to those employees who tested negative initially (Figure 1-6). Employment termination rates during this same period of time were 77 percent higher (23.64% versus 13.35%) for those employees testing positive when compared to those testing negative at the time of preemployment testing (see Figure 1-7).

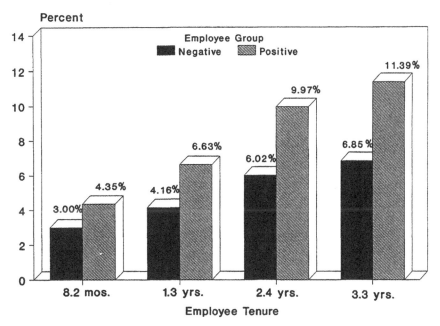

FIGURE 1-6. Absenteeism: U.S. Postal Service study. *Source:* U.S. Postal Service.[11]

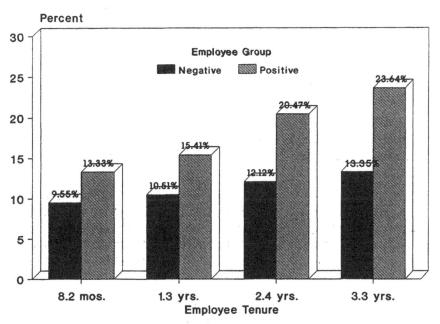

FIGURE 1-7. Firing rates. *Source:* U.S. Postal Service.[11].

In a study conducted at the Georgia Power Company,[12] the relationships between drug use and various measures of workplace performance were examined by comparing records for employees who tested positive in a "for cause" drug testing program with those in a control group made up of individuals who did not use drugs. Individuals testing positive were found to have used almost 8 times the medical benefits (see Figure 1-8) as the control group ($1,377 versus $163) and almost 2.5 times as much as the average worker ($1,377 versus $590). These same individuals were absent (see Figure 1-9) 3.5 times as often as controls (165 hours per year versus 47 hours per year) and 4 times as often as the average worker (165 hours per year versus 41 hours per year).

A 1985 study conducted for DHHS, projected to 1988 for inflation and other factors, estimated that the total economic cost to society of drug abuse was $58 billion a year. Of that amount, $7 billion was estimated to reflect the costs of lost or reduced productivity (morbidity costs); $3 billion was estimated to reflect mortality costs; $2 billion was estimated to represent direct treatment and support costs; $42 billion was estimated to represent other related costs (e.g., direct crime expenditures and productivity losses of addicted individuals who engage in crime as a career rather than legal employment); and, $3 billion was estimated to represent the direct and indirect costs of AIDS associated with intravenous drug users (see Figure 1-10).[13]

COMPONENTS OF A COMPREHENSIVE DRUG-FREE WORKPLACE PROGRAM

E.O. 12564 mandated that each executive federal agency establish and implement a program to achieve a workplace free from drugs, outlining the following elements of a comprehensive drug-free workplace program:

1. A clearly articulated written policy describing the employer's expectations about drug use and consequences of policy violations. It is extremely important that the policy be geared to the needs of the employees and the employer and that it consider the organizational environment (both internal and external). The policy should cover all employees (managers and nonmanagers) within the organization.

2. An EAP to provide confidential problem assessment, counseling, referral to treatment, support and guidance through treatment/rehabilitation, and follow-up support after treatment. EAPs also contribute to articulating

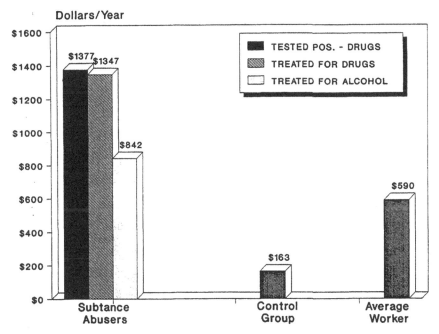

FIGURE 1-8. Medical benefits usage. *Source:* Georgia Power study.[12]

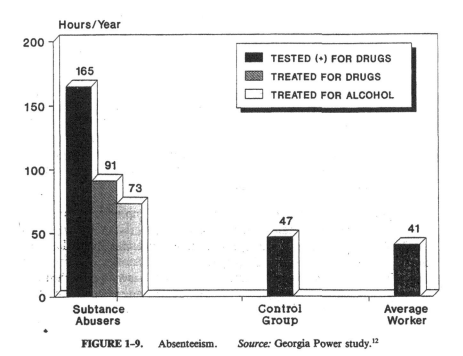

FIGURE 1-9. Absenteeism. *Source:* Georgia Power study.[12]

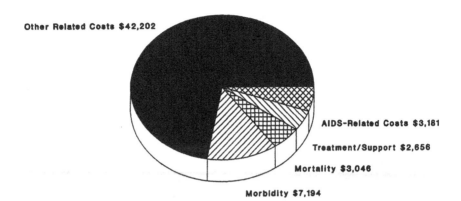

FIGURE 1-10.   Economic costs to society of drug abuse: 1988 estimates. *Source:* D. Rice et al.[13]

company philosophy in a policy statement, forming an advisory committee, assessing the needs of the employee population, providing a comprehensive range of client services, providing employee education, providing supervisor training, establishing linkages within the company and to the outside community, and evaluating the treatment program and other services provided.

3. Supervisor training to orient supervisors to the agency's drug abuse policy, to define the supervisor's responsibility to refer employees when job performance deficits are noted, to provide information on the prevalence and type of problems existing in the work force, to recognize and respond to employees with problems, to provide guidance on how to document changes in employee performance or conduct, and to explain the roles of the EAP and drug testing components.

4. Employee education to describe the signs and symptoms of drug abuse and its effects on performance and to explain the program, including use of the EAP, to employees. All employees, including supervisors, should attend employee education sessions. Employee education can take several forms aside from the "classroom" education sessions, including: scheduling brown-bag luncheon meetings, distributing policy and program materials, and providing information with pay slips, on posters, or through desk-to-desk circulation.

5. Provisions to identify illegal drug abusers including drug testing on a controlled and carefully monitored basis. E.O. 12564 authorizes six types of testing: preemployment, post-accident, reasonable cause, return to duty, random, and voluntary testing. The immediate purpose of testing programs is to protect the health and safety of all employees through the early identification and referral for treatment of employees with substance abuse problems. Drug testing is only one element of a drug-free workplace and should always coexist with programs of drug abuse treatment, prevention, and education.

Although this prototype was developed for the federal agencies' programs, it has critical and necessary elements that also apply to nonfederal employers.

INDUSTRY RESPONSES TO DRUG ABUSE IN THE WORKPLACE

In 1988, the Bureau of Labor Statistics (BLS) conducted the Survey of Employer Anti-Drug Programs[14] to determine the prevalence and nature of antidrug efforts among private nonagricultural employers. BLS found that 3 percent of American industries had a drug testing program, 9 percent had a written policy on drugs, and nearly 7 percent had an EAP program. The survey also showed a tremendous variability by size of industry. Only 2 percent of small businesses (those with 1–49 employees) had drug testing programs, nearly 7 percent had written policies on drugs, and 5 percent had EAP programs. In contrast, approximately 32 percent of large businesses had a drug testing program, nearly 56 percent had a written policy on drugs, and 51 percent had an EAP program.

A repeat survey by BLS in 1990[15] (see Figure 1-11) showed a greater response by industry: Approximately 46 percent of large businesses had drug testing programs, nearly 74 percent had a written policy on drugs, and approximately 79 percent had an EAP program (see Figure 1-12). However, these statistics must be viewed cautiously; small businesses, who represent 80 percent of American businesses, saw relatively minor increases in the elements necessary for a comprehensive drug-free workplace program. Three percent of small businesses had drug testing programs, almost 12 percent had written policies on drugs, and 9 percent had EAP programs. This is particularly troubling when we consider the fact that in 1989 a Gallup poll surveying over one-thousand American workers reported an almost

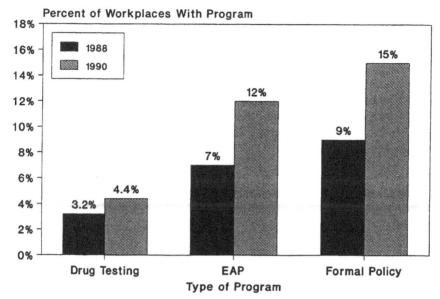

FIGURE 1-11. Employer antidrug programs: 1988 versus 1990. *Source:* H. Hayghe.[15]

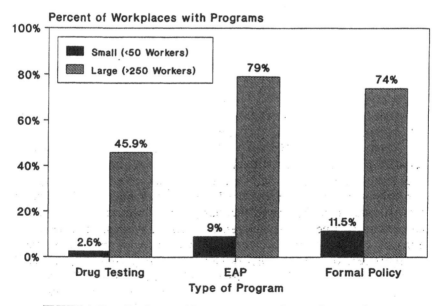

FIGURE 1-12. Employer antidrug programs: small versus large employers. *Source:* H. Hayghe.[15]

unanimous consensus (97 percent) that workplace drug testing is appropriate under certain circumstances and that 85 percent of American workers believe it is also potentially useful in deterring illicit drug use in the United States.

RECENT AND PENDING LEGISLATION

To date, President Bush has presented four national drug control strategies to the Congress and to the American people. The first of these was in September 1989; the second in February 1990; the third in February 1991; and the most recent in January 1992. Taken together, these strategies acknowledge drug abuse as a significant public policy issue associated with rising rates of violent crime, serious damage to the nation's health and economy, and a key tension in international relationships. They propose an integrated series of federal initiatives to reduce the supply of drugs available in the United States and to reduce the demand for drugs by drug users. The fourth national drug control strategy reaffirms the elements of the first three strategies:

> The overall objective of the Strategy is to reduce drug use. This is to be accomplished by reducing both the supply of and demand for drugs. . . . We will keep the pressure on to prevent Americans, especially our children and young adults, from ever starting to use drugs. We must also expand our efforts with adults in the workplace and in the populations that are most resistant to the anti-drug message.[16]

Unlike the previous strategies, the fourth strategy includes provisions to specifically sharpen "the focus on the treatment and prevention of alcohol abuse" and to prevent a resurgence of heroin use as the "availability and purity of heroin increase."[16]

Reducing drugs in the workplace is a key component of the third and fourth strategies. Every employer, large and small, is encouraged to establish a comprehensive workplace drug prevention program. Of particular note is the November 19, 1990 Office of National Control Drug Control Policy publication *Building a Drug-Free Workforce*.* This publication contains model state legislation consisting of four separate bills designed to help public and private employers create incentives for employees to remain drug-free or cease using drugs, and sanctions for those who do not. Its goal is to encourage states to promote comprehensive, consistent drug-free workplace programs, and high-quality drug testing that is both accurate

---

*Building a Drug-Free Workforce can be purchased from the Superintendent of Documents, Government Printing Office, Washington, DC 20402-9325.

and protective of workers' confidentiality. New federal initiatives are anticipated to focus on helping small businesses, which comprise 80 percent of the businesses in the United States, develop drug-free workplace programs. The development of model programs for small businesses is expected to focus on developing ways to make their drug-free workplace programs affordable and cost-effective. For example, one approach has been the development of consortia of small businesses for the purchase of drug testing services, EAP services, and Medical Review Officer (MRO) services. These efforts highlight the administration's conviction that illicit abuse is a national problem affecting all strata of American society.

In line with President Bush's mandate to establish a drug-free America, several pieces of legislation that deal with a drug-free workplace or some component of it have been proposed and, in the case of the Omnibus Transportation Employee Testing Act and the Americans with Disability Act, ratified. A discussion of pertinent legislation follows.

Omnibus Transportation Employee Testing Act

The Omnibus Transportation Employee Testing Act of 1991 requires alcohol and drug testing for employees in safety-sensitive positions in the aviation, rail, truck, bus, and mass transit sectors.[17] The act also acknowledges the critical role of rehabilitation programs, and directs the secretary of DOT to develop regulations regarding the "identification and opportunity for treatment of employees . . . [in safety-sensitive positions] in need of assistance in resolving problems with the use . . . of alcohol or controlled substances." The act requires preemployment, reasonable suspicion, random, and post-accident testing conducted according to the provisions of the NIDA Guidelines and any susbsequent amendments. The act also requires split-specimen collections. (This is a procedure in which the urine specimen is divided into two portions, one of which is analyzed and the other preserved by freezing. In the event of a confirmed positive test, the individual can request that the preserved portion also be analyzed.) The act does not specify the method of alcohol testing, but instead directs the secretary of DOT, in consultation with the secretary of DHHS, to develop regulations for such testing.

The Omnibus Transportation Employee Testing Act amends the Commercial Motor Vehicle Safety Act of 1986, which covers both inter- and intrastate trucking. Therefore, the Omnibus Transportation Employee Testing Act would appear to extend federally mandated drug testing to intrastate trucking operations. (The Federal Highway Administration's drug testing regulations cover only interstate trucking). The Omnibus Transportation Employee Testing Act also provides for federal preemption of conflicting and inconsistent state and local laws, which is particularly noteworthy since approximately 20 state and local jurisdictions have enacted (as of 1991) provisions concerning drug testing that typically limit some aspect of testing.[18] Federal preemption provides certainty and consistency to employers, particularly those engaged in multistate operations.

S. 1903: Quality Assurance in the Private-Sector Drug Testing Act

In 1989, Senators David L. Boren and Orrin G. Hatch introduced Senate Bill 1903 (S. 1903), which proposed creation of uniform standards for drug testing in the private sector.[19] It placed responsibilities on the employer who conducts drug testing to have a written policy, to communicate that policy to employees, and to provide a drug awareness program. Its sponsor, Senator Hatch, said:

> First it will make it clear that American employers currently have the right to implement a drug testing program as part of the drug-free workplace policy. But in affirming that right, the Hatch–Boren bill recognizes that employers are willing to accept the imposition of minimum federal standards in order to guarantee uniformity and predictability with respect to a drug testing program.[20]

S. 1903 proposed that certain standards be required of private sector workplace drug testing programs, including the following:

1. Use of drug testing laboratories certified by a program established by the secretary of DHHS for analyses of a variety of specimens, for example, urine, blood, breath, hair.
2. A requirement for the employer to put in writing provisions of the drug policy and to conduct a drug awareness program.
3. Allowance of the following types of testing: reasonable cause, post-accident, as part of a scheduled medical examination, and during or after a drug treatment program. Random testing is allowed only for safety-sensitive positions.
4. Employee protections, such as required confirmatory tests; the right to be informed about the employers drug testing program; the right to explain the results of confirmed positive tests (analogous to the MRO function); the right to confidentiality of test results; and the right to be free from retaliation for exercising these rights.
5. Enforcement of laboratory certification provisions by the secretary of

DHHS and enforcement of employee protections by the Department of Labor.

S. 1903 would also preempt conflicting or inconsistent state and local laws.

H.R. 33: The Drug Testing Quality Act

In 1990, Representatives Thomas J. Bliley and John D. Dingell introduced House Resolution 33 (H.R. 33). In its original form, H.R. 33 essentially required that all private-sector drug testing programs use NIDA-certified laboratories.[21] H.R. 33 was originally passed as an amendment to the Crime Bill, but was withdrawn before final passage of that bill. H.R. 33 was reintroduced with some modifications in 1991.[22]

H.R. 33 proposed that certain standards be required of private-sector workplace drug testing programs, including the following:

1. Use of drug testing laboratories certified by standards that are the same as those in federally mandated programs.
2. Allowance of testing for the "NIDA 5" drugs and drug classes, barbiturates listed under Schedules I and II of the Controlled Substances Act, oxazepam and alprazolam of the benzodiazepine class of drugs, and anabolic steroids. H.R. 33 also allows individuals to petition the Secretary of DHHS to develop standards to test for additional drugs and drug classes. The secretary may grant to athletic testing programs exceptions for drug testing standards. H.R. 33 also provides for an every-other-year review of the advisability of, and need for, incorporating additional drugs and drug classes.
3. Mandating of standardized procedures for urine specimen collection, storage, and security.
4. Employee protections, such as required confirmatory tests and review of laboratory-positive results by an MRO.
5. Civil and criminal penalties and a private right of action for: violation of federal standards; testing for drugs not specifically included in the laboratory's certification; testing for medical conditions without the subject's consent; altering or falsely reporting test results or adulterating specimens; taking action against an employee prior to gas chromatography/mass spectrometry testing; knowingly violating the requirements of H.R. 33.

H.R. 33 would also preempt conflicting or inconsistent state and local laws.

When testing for commonly used legal drugs—for example, as proposed by H.R. 33—the purpose goes beyond deterring people from illegal drug use. If the person has a valid prescription, the question often becomes whether or not the medication impairs the person's fitness for duty. This is also true of alcohol testing, since the mere presence of ethanol does not equate with abuse. There is an increased emphasis, reflected in recent and proposed legislation, on determining fitness for duty, rather than merely detecting the presence or absence of illegal substances.

Americans with Disabilities Act

The provisions of the Americans with Disabilities Act (ADA)[23] take effect for employers with more than 25 employees in July 1992, and for employers with more than 15 employees in July 1994. The ADA expands the definition of disability to cover individuals who are suspected of having a disability. According to § 3(2) of the ADA, *disability* means (1) a physical or mental impairment that substantially limits one or more major life activities of an individual, (2) a record of such impairment, or (3) being regarded as having such an impairment. The ADA prohibits employers, employment agencies, labor organizations, and joint labor-management committees from discriminating against any qualified individual with a disability because of such disability in regard to job application procedures, hiring or discharge, compensation, advancement, training, or any other term or condition of employment.

In recognition of the widespread concern over illict drugs in the workplace, the ADA contains a separate section (§ 104) dealing with drug abuse and drug testing. The ADA states that the phrase "qualified individual with a disability" does not include any employee or applicant who currently uses illegal drugs. Employers are specifically permitted by § 104(c) of the ADA to:

1. Prohibit the use of alcohol or illegal drugs at the workplace by all employees.
2. Prohibit employees from being under the influence of alcohol or illegal drugs at the workplace.
3. Require employees to follow the requirements of the Drug-Free Workplace Act of 1988.
4. Require employees to meet the job-related requirements established by the U.S. DOT, DOD, and NRC with respect to drugs and alcohol.
5. Hold a drug user or alcoholic to the same qualification standards for employment or job performance or behavior to which it holds other individuals, even if unsatisfactory performance or behavior is related to the drug use or alcoholism of the individual.

The ADA does not impact an employer's ability to perform drug testing for applicants or employees. A drug test is not considered a "medical examination," and thus is not subject to the same restrictions as other employment physicals. Also, § 104(d)(2) of the ADA specifically permits employers to conduct drug testing of job applicants and employees and make employment decisions based on the results of those tests. Finally, the act specifically permits employers to "adopt or administer reasonable policies or procedures, including but not limited to drug testing" [§ 104(b)(3)] to ensure that individuals undergoing, or who have completed, drug rehabilitation treatment are not currently using illegal drugs.

SOURCES FOR ADDITIONAL INFORMATION

The NIDA Toll-Free Workplace Help Line is a telephone consultation service for employers and labor representatives on the topic of developing a drug-free workplace policy. The phone number is 1-800-843-4971. The help line operates from 9 A.M to 8 P.M eastern standard time, Monday through Friday.

The National Clearinghouse for Alcohol and Drug Information (NCADI) provides model policies, free publications, and literature searches and lends videotapes on drugs in the workplace. NCADI can be contacted in writing (P.O. Box 2345, Rockville, Md. 20852) or by calling 1-800-729-6686, or in the D.C metro area call 301-468-2600.

The NIDA Toll-Free Drug Abuse Information and Treatment Hot Line provides drug-related information to callers, assists them in placing drug abusers in treatment programs, and familiarizes them with community programs and support groups in their areas. The phone number is 1-800-662-4357. The hot line operates from 9 A.M to 3 A.M. (EST), Monday through Friday, and from noon to 3 A.M., Saturday and Sunday.

The National Association of State Alcohol and Drug Abuse Directors (NASADAD) (202-783-6868) will provide the name and phone number of state alcohol/drug abuse directors. The state director will provide information pertaining to activities and services in that state.

The Employee Assistance Professionals Association (EAPA) provides EAP-related information to callers on the national, regional, and chapter levels. EAPA publishes EAP materials, and trains and certifies EAPs.

The U.S. Department of Transportation's Anti-Drug Information Center offers information on drug testing, policy, and regulations covering transportation workers. This service can be accessed by telephone (1-800-CAL-DRUG) and by modem (1-800-225-3804).

SUMMARY

Drug abuse is a pervasive public health problem. The late 1980s saw unprecedented growth in awareness and concern over drug use and its toll on health and workplace productivity. Federal regulations mandate a comprehensive approach to deterring drug abuse in the federal work force and in certain private-sector areas. Future regulations will probably further standardize drug abuse programs and, in particular, drug testing procedures in the private sector. They will also likely focus more attention on fitness for duty and licit drugs, including alcohol.

Acknowledgement

The authors would like to thank Judith T. Galloway, J.D., for her contributions to this manuscript.

References

1. Reagan R. Executive Order 12564: Drug-free federal workplace. *Fed. Reg.* 1986;51(Sept. 27):32,889-32,893.
2. Supplemental Appropriations Act of 1987. Section 503, pub law 100-71, 101st Congress, July 11, 1987.
3. Department of Health and Human Services (DHHS). Mandatory guidelines for federal workplace drug testing programs. *Fed. Reg.* 1988;53(April. 11):11,970-11,989.
4. U.S. Department of Transportation. Procedures for transportation workplace drug testing programs. *Fed. Reg.* 1989;54(Dec. 1):49,854-49,884.
5. Drug-Free Workplace Act of 1988. Title V, Subtitle D, Sections 5152 and 5153, pub law 100-690, 100th Congress, November 18, 1988.
6. U.S. Department of Defense. Drug-Free work force. *Fed. Reg.* 1988;53(Sept. 28):37,763-37,765.
7. Nuclear Regulatory Commission. Fitness-for-duty programs. *Fed. Reg.* 1989;54(June 7):24,468-24,508.
8. National Institute on Drug Abuse. *Summary of Findings from the 1991 National Household Survey on Drug Abuse.* Rockville, Md.: NIDA Capsule C-86-13, rev. December 19, 1991.
9. National Institute on Drug Abuse. *National Household Survey on Drug Abuse: Population Estimates 1990.* Rockville, Md.: U.S. DHHS publication no. (ADM)91-1732; 1991.
10. National Institute on Drug Abuse. *National Household Survey on Drug Abuse: Population Estimates 1988.* Rockville, Md.: U.S. DHHS publication no. (ADM)89-1636; 1989.
11. U.S. Postal Service. Utility analysis of preemployment drug testing as a selec-

tion device. Paper prepared by the Personnel Research and Development Branch, Office of Selection and Evaluation; June 1991.
12. Winkler, H., and J. Sheridan. An analysis of workplace behaviors of substance users. Paper presented at Drugs in the Workplace: Research and Evaluation Data. Rockville, Md.: National Institute on Drug Abuse; September 1989.
13. Rice D., S. Kelwan, L.S. Miller, and S. Dunmeyer. *The Economic Costs of Alcohol and Drug Abuse and Mental Illness: 1985.* Washington, D.C.: U.S. DHHS publication no. (ADM)90-1694;1990:26.
14. Bureau of Labor Statistics. *Survey of Employer Anti-Drug Programs.* Washington, D.C.: U.S. Department of Labor Report 760; January 1989.
15. Hayghe, H. Anti-drug programs in the workplace: Are they here to stay? *Monthly Labor Rev.* 1991;114:26-29.
16. The White House. *National Drug Control Strategy.* January 1992.
17. Omnibus Transportation Employee Testing Act of 1991. Pub law 102-143, 102nd Congress, October 28, 1991.
18. *A Guide to State Drug Testing Laws & Legislation.* Washington, D.C.: Institute for a Drug-Free Workplace; 1991.
19. Quality Assurance in the Private Sector Drug Testing Act of 1989. S. 1903, 101st Congress, 1st sess., sec. 1, 1989.
20. Hatch, O. Quality assurance in the Private Sector Drug Testing Act. *Congressional Rec.* 1989;135(Nov. 17):16,033-16,038.
21. Standards for Certification of Laboratories Engaged in Drug Testing. H.R. 33, 101st Congress, 1st sess., sec. 1, 1989.
22. Drug Testing Quality Act of 1991. H.R. 33, 102nd Congress, 1st sess., sec. 1, 1991.
23. Americans with Disabilities Act of 1990. Pub law 101-336, 101st Congress, July 26, 1990.

2

Drug Testing in the Workplace

Robert Willette, Ph.D., and
Leo Kadehjian, Ph.D.

HISTORY OF WORKPLACE DRUG TESTING

Drug testing emerged in the 1960s as a diagnostic tool to assist in the evaluation and treatment of drug users. The basic principle of drug testing is to detect the presence of specific drugs or their metabolites. The results of such tests serve to identify the drug(s) taken by the individual, and, in many drug treatment programs, to deter the individual from continued use of drugs. Urine is the preferred specimen for testing because it is easily obtained and contains high concentrations of drugs and drug metabolites. Generally, the drug testing methods employed during the 1960s and 1970s were moderately specific (i.e., correctly identified drugs) but relatively insensitive (i.e., did not detect low concentrations of drugs). This disadvantage was countered by conducting tests more frequently, without notice, and in a random manner.

In the late 1960s and throughout the 1970s, intensive research and development efforts provided new and better tools for conducting drug tests and a wealth of data to help in the interpretation of their results. The federal government, primarily through the National Institute on Drug Abuse (NIDA), sponsored studies on drug metabolism, disposition, and excretion.

Support was provided for the development of analytical methods for detecting and identifying drugs and their metabolites, including pioneering efforts with gas chromatography-mass spectrometry (GC/MS).

Concurrently, private companies began producing immunoassays for drugs of abuse. These assays were significantly better and more efficient than older methods for testing urine specimens. The first immunoassays were relatively sensitive and inexpensive, and could process large numbers of specimens. In the late 1970s, portable immunoassay methods became available, making on-site testing possible at treatment and criminal justice settings. By the early 1980s, immunoassays were available for 10 different drug types.

By 1980, several laboratories had emerged that offered drug testing services for large numbers of specimens at reasonable costs. Better confirmation tests were developed. In the early 1980s, laboratories relied primarily on gas chromatography to identify specific drugs and their metabolites. In the mid-1980s, testing was improved by adding mass spectrometry analysis, which was made possible by the advent of compact, easy-to-use, less expensive instruments.

Concern about the accuracy of drug testing grew in the early 1970s. More laboratories began performing analyses when drug testing became required in federally funded, federally regulated drug treatment programs. Although results were intended to diagnose and assist in the counseling of patients, positive drug results could have adverse consequences, such as expulsion from the treatment program. In order to protect patients, a drug testing laboratory monitoring program was initiated in 1972. Although limited in its regulatory powers, this NIDA-sponsored program, conducted by the Centers for Disease Control (CDC), provided some guidance and assurances about the reliability of drug testing.[1] This program ended in 1981, leaving no government oversight of private drug testing laboratories until 1988.

Despite acknowledged limitations, the technological advances described above made drug testing more affordable and reliable and played a major role in extending its use from the drug treatment setting in the 1960s to workplace testing in the 1970s and 1980s.

Private-Sector Testing

There is no accurate record of who started drug testing among private employers. It was fairly common for many employers to have policies that addressed intoxication—particularly, alcohol intoxication—on the job. Some interstate bus and trucking companies included drug tests as part of their biannual medical examinations required under Subpart E of the Federal Highway Administration (FHWA) regulations.[2] Although Subpart E does not specifically require drug testing, it does state that use of certain addictive drugs may classify a driver as medically unqualified to operate a vehicle. One of the earliest drug testing arbitrations occurred in the early 1980s when the Greyhound bus lines began testing its drivers for marijuana without advance notice.

Many other companies instituted preemployment testing programs during this period. In 1983, the Edison Electric Institute issued guidelines for operators of nuclear and other power plants on drug abuse and drug testing programs.[3] This first industry-wide voluntary program was modeled closely after the military programs. Other industries followed the lead. Notably, petroleum, chemical, and certain transportation-related companies were amongst the earliest. In transportation, the railroads, under Federal Railroad Administration (FRA) regulations, were the most aggressive. By the mid-1980s, nearly half of the major companies in the United States were conducting preemployment drug testing and a growing number were conducting reasonable cause and postrehabilitation testing. A few started doing random testing. The number of companies continued to grow over the second half of that decade.

One of the problems that remained with these pioneering drug testing programs was the lack of government oversight or certification of drug testing laboratories. The Food and Drug Administration (FDA) regulated commercially sold diagnostic kits, such as the immunoassay tests. However, there was no such registration or control over the people using them. Many, but not all, of the laboratories offering workplace drug testing services adhered to generally accepted practices in the handling and testing of specimens.

During the early 1980s, employers had little guidance in selecting and monitoring drug testing laboratories. Many learned the hard way, either in arbitrations or in court. For example, the navy's program got off to a shaky start because its laboratories were not adequately prepared for the sudden increase in the number of specimens and were not sufficiently trained in forensic principles. Until 1981, these were primarily clinical laboratories, doing drug testing as part of their routine toxicological testing. The nature of forensic testing—in which specimens constitute evidence, and each specimen is accounted for through "chain of custody" procedures*—was foreign to clinical settings. It took an unfortunate number of errors and poor laboratory practices to bring order to the drug testing system.

Federally Regulated Drug Testing

The Military

The screening of individuals without outward signs of drug use started in the U.S military forces during the Vietnam conflict in the late 1960s. The

---

*Chain of custody procedures account for the integrity of each specimen by tracking its handling and storage from point of specimen collection to the final disposition of the specimen.

ready availability of heroin and hashish tempted many service personnel. The military established drug screening programs to test its servicemen (and women), especially before they returned home from Vietnam. The results of testing were intended only to direct servicemen into treatment. At the time, various administrative and court decisions restricted the services from taking disciplinary action.

This treatment-oriented approach to drug abuse in the military failed. In a worldwide survey conducted by the Department of Defense (DOD) in 1980, the services were shocked by the prevalence of drug abuse among their personnel.[4] For example, 48 percent of enlisted navy personnel admitted to marijuana use within a 30-day period. Over 20 percent admitted such use in each of the other services. The navy conducted an anonymous, random urinalysis survey in 1981 to investigate these findings: 47 percent of specimens from enlisted personnel tested positive, which prompted the navy to initiate an aggressive drug control program that included random drug testing and the potential for disciplinary consequences.

The key to this change in policy toward drug abusers in the military was a memorandum signed by then Assistant Secretary of Defense Frank Carlucci authorizing the military to use drug testing results for discipline, including discharge. In order to stop and deter drug abuse, DOD issued specific directives on how to conduct testing and under what circumstances test results could be used for discipline. The procedures were carefully designed to ensure the integrity of the system and to stay within the strict legal boundaries imposed by the Military Rules of Justice, the U.S. Constitution, and the courts. For example, only random and probable cause test results could be used for disciplinary purposes.

The navy's aggressive approach over the 10 years of its War on Drugs has reduced detected drug use to below 2 percent.[5] A study conducted for the navy in 1984 revealed that, of the measures the navy employed to address drug abuse, the most effective deterrent was random testing, coupled with the possibility of discipline.[6] Thus, drug testing was established as an important tool to identify abusers and to deter drug use.

The primary regulation governing clinical laboratories, the Clinical Laboratories Improvement Act of 1967, had no specific provisions for drug testing.[7] A few states—most notably New York, Pennsylvania, and California—had their own laboratory licensing programs. The CDC conducted a laboratory monitoring program sponsored by NIDA that was limited to laboratories serving drug treatment programs that NIDA funded or that were licensed by the FDA to dispense methadone, a drug used as a substitute narcotic to treat heroin addiction; this program ended in 1988, when funding was shifted to the states. At about the same time, massive problems were discovered in one of the navy's drug testing laboratories. Ironically, one federal agency discontinued its monitoring of drug testing laboratories while another agency intensified its monitoring.

The Armed Forces Institute of Pathology (AFIP) has monitored the military's drug testing laboratories since the early 1970s by submitting performance test (PT) samples—that is, urine samples to which known amounts of drugs are added—to test each laboratory's analytic accuracy and precision. Unfortunately, when the navy rapidly accelerated its testing in 1982, a large backlog of specimens delayed the processing of the PT samples and allowed one of the navy drug testing laboratories to produce thousands of results of questionable validity. Similar problems were found at other military laboratories.

In order to improve the military's drug testing, the White House directed DOD and NIDA to bring together noted forensic experts and laboratory representatives of the services. The message was clear: In order to provide the protection necessary to sustain disciplinary action on the basis of a single urine specimen collected without suspicion, the test procedures had to meet forensic standards. The battle cry became: "The results must stand up in court." In other words, one would have to be able to prove beyond a reasonable doubt that the specimen came from that specific individual, that no one had an opportunity to tamper with or substitute the specimen, and that any drug analyte found was actually present.

The AFIP program grew stronger. Each branch of the military adopted stringent protocols and system-wide standard operating procedures (SOP). Highly trained personnel were assigned to the laboratories. Most importantly, the laboratories were opened to external scrutiny and periodic inspections by civilian experts. Several thousand positive results that had been previously declared "positive" were invalidated because of inadequate documentation regarding specimen collections and analyses. The records of those affected were expunged. Thus, DOD-regulated drug testing became more reliable and recognized as the "gold standard."

The NIDA Guidelines

Starting late in 1985, a few federal agencies, including the U.S. Customs Service and the Drug Enforcement Administration, began to evaluate drug testing as a means to detect and deter drug use among their employees. These agencies had strategic roles in the government's war on drugs and needed assurance that their agents were drug-free. Testing began in early 1986 and was limited to persons entering "sensitive" positions. In September of that year, President Reagan issued Executive Order (E.O.) 12564, which directed all federal agencies to implement drug testing programs for federal employees holding safety- or security-sensitive positions.[8] Testing would include preemployment, post-accident, reasonable cause, and random. (These types of tests are explained later in this chapter.) The random testing component was to be limited to the five more widely abused drugs: marijuana, cocaine, amphetamines, opiates, and phencyclidine (PCP). In reasonable cause and post-accident testing, any drug classified in Schedules I and II of the Controlled Substances Act—a federal law governing the distribution of depressant and stimulant drugs—could be included.

In order to establish standards for this testing, the president directed the Department of Health and Human Services (DHHS) to develop scientific and technical guidelines for all aspects of the testing procedures, that is, collection of specimens, chain of custody, laboratory accreditation, and medical review of results. The Supplemental Appropriations Act of 1987 reinforced E.O. 12564 by requiring a more formal laboratory certification program and the publication of the guidelines in the *Federal Register*.[9] DHHS published its "Mandatory Guidelines for Federal Workplace Drug Testing Programs" in proposed form in August 1987,[10] and in final form in April 1988.[11] These regulations, often referred to (incorrectly) as the NIDA Guidelines, for the National Institute on Drug Abuse, contained procedures that were mandatory for any federal agency that conducted drug testing.

The NIDA Guidelines were the template for subsequent regulations by the Department of Transportation (DOT) and the Nuclear Regulatory Commission (NRC), and were adopted by many private-sector programs. The NIDA Guidelines established detailed procedures for specimen collection and handling, introduced the concept of the Medical Review Officer (a physician who interprets drug test results), and introduced a comprehensive certification program for laboratories that performed drug testing. NIDA was given responsibility for implementing and overseeing the certification program. In December 1988, there were 10 NIDA-certified laboratories; by 1991, over 80 laboratories were certified.

The NIDA Guidelines have been criticized for limiting testing to only five drug classes, and for their required cutoffs (concentrations at or above which a test is called positive), reporting requirements, and laboratory certification standards. Some private-sector employers voluntarily adopted the NIDA Guidelines into their programs, but others felt the NIDA Guidelines were too restrictive and would undermine their existing programs. To provide the private sector with an opportunity to discuss these differences and to recommend modifications to the NIDA Guidelines, NIDA held a consensus conference on *Technical, Scientific and Procedural Issues of Employee Drug Testing* in November 1989. From this conference came the "Consensus Report" of recommendations for modifying the NIDA Guidelines.[12] Revisions to the guidelines were anticipated based on these recommendations.

Department of Transportation Procedures

The DOT published its "Procedures for Transportation Workplace Drug Testing Programs" in interim form in November 1988,[13] and in final form in December 1989.[14] They were a set of "procedures and safeguards" for DOT-regulated drug testing programs. Separately, in November 1988, six DOT modal agencies published regulations that incorporated, by reference, the DOT Procedures, and established policy requirements for drug testing conducted under the authority of each agency. These six DOT modal agencies are the Federal Aviation Administration, the Federal Railroad Administration, the Federal Highway Administration, the Coast Guard (for marine operations), the Research and Special Projects Administration (for pipeline operations), and the Urban Mass Transit Administration. The drug testing requirements of the first five modes have taken effect. The Urban Mass Transit Administration (UMTA, now named the Federal Transit Administration, FTA) program was initially suspended by the U.S. Court of Appeals, which ruled that UMTA did not have authority in this matter. In 1991, legislation was passed to reinstate FTA's drug testing requirements.

The DOT Procedures require collection, analysis, medical review, and reporting procedures that are similar to those of the NIDA Guidelines. Most importantly, all DOT-mandated testing must be done in NIDA-certified laboratories.

Department of Defense Regulations

The DOD issued an interim rule for a "Drug-Free Work Force" that would require each DOD contractor to implement drug testing as part of a drug-free workplace program.[15] The interim rule did not elaborate on the nature and extent of drug testing, but clearly provided DOD contractors with the authority to implement such programs. DOD's final rule, published in November 1991,[16] applies to DOD contractors who have employees working in sensitive positions. Among other things, the final rule requires random testing and procedures that comply with the NIDA Guidelines, and supersedes any conflicting state and local laws.

Nuclear Regulatory Commission Regulations

The nuclear power industry had implemented rather aggressive drug testing programs in the early 1980s. In order to standardize and strengthen these early voluntary programs, in 1989 the NRC issued drug testing regulations.[17] The NRC's technical procedures follow those of the NIDA Guidelines, such as requiring the use of NIDA-certified laboratories. NRC also extends the number of drugs that a licensee can test for, requires alcohol testing, permits the use of lower detection levels, authorizes the use of on-site testing to screen specimens before they are sent to a certified laboratory for confirmation, and permits limited employer access to screening test results.

Department of Energy Regulations

The Department of Energy (DOE) owns and contracts for the operation of several facilities that conduct research on and produce nuclear materials. In 1991, the DOE issued a proposed rule, "Workplace Substance Abuse Programs at DOE Facilities."[18] The proposed rule would require procedures that are consistent with the NIDA Guidelines, and analyses by NIDA-certified laboratories. Testing for alcohol was included in an earlier proposed rule for DOE's "Personnel Security Assurance Program."[19]

State and Municipal Drug Testing Laws

Some states and a few localities, such as San Francisco, have also passed laws regarding drug testing.[20,21] These actions have generally been taken to protect individuals from invasions of privacy or denial of their due process. They primarily address laboratory testing procedures and, less often, employer policy issues. For example, some laws require use of licensed or certified—for example, NIDA-certified—laboratories. Some states have mandated random testing of truck drivers and allow the random testing of persons in safety-sensitive positions, such as police and fire fighters. This array of different rules has caused considerable confusion and nonuniform treatment of employees working for national companies. For example, a company may be able to randomly test employees in one state, but not in another.

The U.S. Congress has considered laws that would impose certain federal requirements on all drug testing, thereby encouraging uniformity. Congressmen Dingell and Bliley introduced a bill, H.R. 33, that would require all employers to use federally certified laboratories, and would require medical review of each positive result.[22] Senators Hatch and Boren have introduced a bill, S. 2008, that would accomplish many of the same objectives as H.R. 33, but would reduce the certification requirements, would allow on-site testing (banned in H.R. 33), would expand the list of drugs that may be tested for, and would allow nurses and "other trained individuals," as well as physicians, to serve as Medical Review Officers.[23] Congress will likely pass legislation, perhaps one of these bills, that standardizes testing across the country.

EVALUATING THE NEED FOR A DRUG TESTING PROGRAM

A number of studies have demonstrated the utility and cost-effectiveness of drug testing programs.[24-28] Employers should nevertheless determine their objectives and consider whether drug testing will help accomplish them. The costs of drug testing—the economic expense and the effects on employee morale—should be weighed against its potential benefits.

Assessing Workplace Drug Abuse

The employer should first assess the impact of drugs on the health, safety, and productivity of the workers. These data can be the most persuasive in justifying drug testing to the employees, unions, and, if necessary, to the courts. Drug testing programs for airline employees, police, pipeline workers, Department of Education employees, U.S. Customs agents, and others have, nevertheless, been legally upheld without evidence of a workplace drug abuse problem.

A 1990 survey found that, among 18- to 34-year-old full-time employed Americans, 24.4 percent reported illicit drug use in the past year, and 10.5 percent reported illicit drug use in the past month.[29] In a 1989 Supreme Court decision upholding drug testing for the U.S. Customs Service, Justice Kennedy wrote: "There is little reason to believe that American workplaces are immune to the pervasive social problem of drug abuse."[30] Prevalence studies of drug abuse in the workplace have corroborated this opinion.

For example, a 1986 study of tractor-trailer drivers voluntarily tested at a roadside weigh station showed an overall 29 percent positive rate for drugs or alcohol (12% of the drivers declined to participate).[31] The Supreme Court found such data supportive of the need to test transportation workers for drugs.[32] A 1988 large, nationwide survey of drug testing conducted by private, nonagricultural employers found overall positive rates of 11.9 percent for applicants and 8.8 percent for current employees (see Table 2-1).

SmithKline Beecham Clinical Laboratories reported that, in 1991, 8.8 percent of more than 2 million drug tests it analyzed were laboratory-positive, down from 11 percent in 1990, 12.7 percent in 1989, 13.6 percent in 1988, and 18.1 percent in 1987. Of SmithKline's 1991 positive results, 34.6 percent were for marijuana, and 29 percent were for cocaine.[34]

Legitimate substance use—for example, alcohol and certain prescription medications—can also impair performance. The cost to industry from alcohol abuse is high, despite a decline in per capita alcohol consumption since 1970. In 1983, alcohol abuse cost society an estimated $117 billion; 61 per-

TABLE 2-1 Drug Testing Results for Job Applicants and Current Employees

Industry	Applicant Positive (%)	Employee Positive (%)
Mining	12.7	6.1
Construction	11.9	12.0
Manufacturing		
Durable	11.2	12.1
Nondurable	12.7	8.9
Transportation	9.9	5.6
Communications/Public Utilities	5.5	7.8
Wholesale Trade	17.4	20.2
Retail Trade	24.4	18.8
Finance, Insurance, Real Estate	6.7	0
Other Services	9.9	3.1
Overall	11.9	8.8

Source: Adapted from *Survey of Employer Anti-Drug Programs.*[33]

cent of this amount was due to lost employment and reduced productivity. In 1990, alcohol abuse cost society an estimated $128 billion, approximately half of this amount due to lost employment and reduced productivity.[35] Nearly one-half of all traffic fatalities are alcohol-related. Employers concerned with escalating health care costs should note that health-care costs of untreated alcoholics are twice those of non-alcoholics. By comparison, illicit drug use costs businesses an estimated $3 billion from absenteeism, turnover, workmen's compensation, medical claims, and decreased productivity.[36]

Effect of Drug Use on Workplace Performance

A number of studies have demonstrated decrements in psychomotor and cognitive functions after drug use.[37,38,39] Of the few studies that have addressed drug testing and workplace performance, most have looked at pre-employment, rather than in-service, testing. These studies have looked for an association between preemployment drug test results and employment outcomes.

In the following three studies, drug test results were used only for research purposes, and were kept confidential from the employer:

- In one hospital employee study,[40] 180 employees underwent preemployment drug tests; 22 of them tested positive. The study found no statistically significant association between drug test results and measures of job performance and retention. However, this study's ability to detect significant results was limited by the small size of the group.

- Of 4,375 postal service employees, nationwide, that underwent preemployment drug tests, 8.4 percent tested positive.[41,42] The employee group that tested positive had 41 percent higher absenteeism, 1.5 times more involuntary turnover, and 1.7 times greater likelihood of quitting than the group that had negative drug tests. The cost savings that could be obtained by testing was an estimated $3.1 million in the first year, and $17 million over 3 years.

- Of 2,537 employees of the Boston Postal Service that underwent preemployment drug tests, 12.2 percent tested positive.[43] The employees were followed for an average of 406 days. The employee group that tested positive had a higher rate of several adverse job outcomes than the group that tested negative (see Table 2-2). The degree of risk was less than that seen in other studies. An accompanying editorial suggests that, had this study better distinguished between chronic and casual drug abusers, it might have found higher rates of adverse employment outcome in the chronic abusers.[44]

In a study of U.S. Navy Recruits,[45] 500 recruits who tested positive for marijuana when inducted were compared with a matched group of 500 recruits who tested negative. A marijuana-positive drug test at induction led to warnings, counseling, and in some cases surveillance, but did not lead to dismissal. A subsequent marijuana-positive result, however, was cause for dismissal. After two and a half years, there was greater retention (81%) in the marijuana-negative group than in the marijuana-positive group (57%).

TABLE 2-2 Relative Risks of Adverse Employment Outcomes in Employees Testing Positive

Employment Outcome	Marijuana-Positive	Cocaine-Positive
Turnover	1.56	1.15
Accidents	1.55	1.59
Injuries	1.85	1.85
Disciplinary Action	1.55	1.40
Absenteeism	1.56	2.37

Source: Data from Zwerling, Ryan, and Orav.[43]

Note: $\text{Relative risk} = \dfrac{\text{Rate among employees testing positive}}{\text{Rate among employees testing negative}}$

In the marijuana-positive group, 14 percent left the service for drug or alcohol problems, and 1 percent of those in the marijuana-negative group left. In the marijuana-positive group, 21 percent were discharged for other behavioral or performance reasons, and 8 percent of the marijuana-negative group were discharged. These data indicate an increased likelihood of adverse employment outcomes for people with positive preemployment drug tests.

How Employers Have Responded

Drug testing is one of many available methods that can be used to reduce drug abuse in the workplace. But, clearly, testing is the method most on employers' minds today. The following surveys have looked at industry use of drug testing:

- A 1988 nationwide Bureau of Labor Statistics study[33] found that 19.6 percent of employees worked in an establishment that had a drug testing program. Larger employers were more likely than smaller employers to have a program; for example, 59.8% of employers with 5,000 or more employees had drug testing programs, decreasing to fewer employers with programs, proportional to the number of employees, with a low of 0.8 percent for employers with 1 to 9 workers that had drug testing programs.

- A 1988 Gallup poll[46] of 1,000 companies found that 42 percent of companies with 1,000 to 5,000 employees and 60 percent of companies with over 5,000 employees had drug testing programs. Of the large companies with drug testing programs, 68 percent reported having dealt with incidents of drug use within the past year, and 54 percent said that safety was the main reason they conducted testing. Among the benefits of drug testing, 26 percent claimed better-quality applicants, 15 percent claimed reduced accidents, and 14 percent claimed increased productivity.

- A 1991 American Management Association study[47] of 1,633 firms found that 66 percent of respondents engaged in drug testing.

These data indicate that workplace drug testing is a significant part of industry's response to drug abuse. A 1989 Gallup poll of 1,007 workers[48] found that most supported drug testing:

- 41 percent acknowledged that drug use seriously affects job performance;
- 32 percent knew of illegal drugs being sold at their workplace;

- 82 percent supported company antidrug policies, including disciplinary actions;
- 97 percent found testing appropriate;
- 26 percent found testing a necessity;
- 53 percent supported random testing (30% opposed random testing).

Purpose of Testing

Each employer should define the purpose that drug testing would serve within the context of that firm's needs. Drug testing is most often performed to:

- Protect workplace and public safety;
- Maintain productivity by avoiding excess absenteeism, tardiness, and impaired workplace performance;
- Protect the security of property and confidential information;
- Avoid excessive medical and unemployment compensation costs;
- Avoid potential liability from the acts of impaired employees;
- Comply with regulations, for example, DOT, NRC, and DOD.

Most employers that conduct drug testing cite safety and productivity concerns. Testing can also foster public trust, especially where public safety is an issue.

Observed Benefits of Drug Testing Programs

In the following studies, workplace drug testing programs produced dramatically beneficial results:

- The U.S. Navy instituted widespread drug testing after the 1981 airplane crash on the aircraft carrier *Nimitz*. At the time, the navy found that 41 percent of those under age 25 tested positive.[4] After 10 years of random drug testing, the rate of positive results dropped below 2 percent. The navy found that testing, coupled with possible punitive consequences, was the most important factor in this decline.[5]
- A study at the Utah Power and Light Company found that employees testing positive for drugs took 35 percent more sick leave, took 240 percent more unexcused absences, and had five times more reportable vehicular accidents than employees testing negative. A cost-benefit analysis found substantial potential savings from the company's drug testing program.[27]
- Southern Pacific Transportation Company initiated drug testing in 1984.

The rate of employees testing positive fell from 23 percent in 1984 to approximately 5 percent in 1988. Between 1983 and 1988, personal injuries per man-hour worked dropped by a factor of 3, train accidents attributable to human error dropped by a factor of 10, and the costs of such accidents dropped by a factor of 5.[28]

BASIC ELEMENTS OF A DRUG TESTING POLICY

The guidance in this section reflects regulations and major court decisions through summer 1991. The practice of drug testing is rapidly evolving. While this section cannot serve as a substitute for professional legal advice, it intends to identify areas for further consultation with legal counsel.

A comprehensive workplace substance abuse policy includes the following elements:

- Statement of need, for example, a documented drug problem, or concerns about productivity, employee health and fitness, or employee or public safety;
- Statement of purpose, for example, employee health, safety, and productivity, and company competitiveness;
- Employer responsibilities;
- Employee responsibilities;
- Disciplinary actions.

The employer should form a committee to help develop a substance abuse policy. This committee should include representatives from all groups within the company. These may include medical staff, human resources personnel, safety and security representatives, employee assistance program personnel, legal counsel, and union and other employee representatives, where available. Consultants with expertise in policy development and implementation can also assist. A 1988 survey found that in 93 percent of large companies, personnel or human resources departments were "very involved" in implementing the drug testing program; in 62 percent, medical departments were "very involved."[46]

The federal government has provided guidance to federal agencies for implementing a drug-free work place[49] as called for by E.O. 12564. This guidance includes requirements for a policy statement, an Employee Assistance Program (EAP), supervisory training, availability of self-referral, and drug testing for employees in sensitive positions as well as the availability of voluntary drug testing. The guidance authorizes (but does not require) the following types of testing:

- Applicant testing;
- Reasonable cause testing;
- Testing after accidents and unsafe practices;
- Follow-up testing of individuals who undergo counseling or rehabilitation programs for drug abuse.

NIDA has also published material for private- and public-sector employers to use in developing and implementing workplace substance abuse programs, for example, *Model Plan for a Comprehensive Drug-Free Workplace Program.*[50]

Many legal issues must be addressed in the development of a sound policy and program.[51-55] These include constitutional issues; federal, state, and local laws, regulations, and ordinances; common law practices; and collective bargaining agreements. Because of the legal complexities, it is critical that the drug testing policy committee include legal counsel.

Key Policy Issues

In addition to the key policy issues already discussed, the following should be considered:

Safety and security concerns in defining positions subject to testing. Testing has been clearly upheld by the courts when safety and security concerns are of great importance. Some, but not all, testing programs have been struck down where there is no such concern. Courts have considered safety issues especially important when they affect the public. Safety concerns have been accepted as a rationale for conducting testing of airline workers, chemical weapons plant personnel, private-sector chemical plant personnel, employees possessing firearms, corrections officers, police officers, fire fighters, pipeline workers, railroad personnel, nurses, nuclear power plant personnel, truck drivers, school bus drivers, and public bus drivers. In some cases, courts have distinguished the safety concerns of various job categories within these industries.

Overly broad categories of groups subject to testing. Some programs have been struck down for overly broad definitions of positions subject to testing. The courts have held that the employer must relate its interest in testing to the practical job duties of the employees it wishes to test. For example, the Supreme Court questioned the testing of all U.S. Customs personnel with security clearances, indicating that this classification was overly broad and might include employees without access to top secret security information.[30] In this case, the testing of all such employees was subsequently upheld.

Applicants' privacy rights relative to current employees. Courts have found that applicants have diminished privacy rights relative to employees. Employees also sacrifice some privacy rights upon entering the workplace. Individual privacy rights must be balanced against the employer's interest in testing. The more compelling the justification for testing, and the less intrusive the testing process, the more likely the program will withstand legal scrutiny.

On-duty versus off-duty use. Drug testing cannot distinguish between off-duty and on-duty drug use. Some might argue that off-duty use is of no concern to the employer. Courts have not found drug testing unconstitutional simply because it does not distinguish on-duty from off-duty use. But, in some public-sector cases, courts have found it necessary for the employer to establish an association between off-duty use and on-duty impairment. Private-sector employers may not be held to the relatively rigorous standards that courts have applied to public-sector programs. The employer should nevertheless attempt to establish a rationale for how off-duty use impacts work performance.

A few studies have addressed prolonged performance-related effects after smoking marijuana. In one study, marijuana smoking produced measurable performance impairment in cognitive and psychomotor tests the morning after smoking.[37] In another study, experienced pilots showed subtle but significant impairment in flight simulator performance 24 hours after smoking a single marijuana cigarette.[38,39] The hangover effects of alcohol intoxication are widely recognized. The best way to address performance-related issues is by specific objective performance criteria. The employer is probably on safest ground if testing is triggered by well-founded, observable job performance decrements or other determinations of reasonable cause.

Who to Test

The first decision in implementing a testing program is determining who to test. Each employer's policy should describe, as specifically as possible, which groups or positions are subject to testing. Union bargaining may be required in making this determination. Federal agency drug testing regulations identify, in relatively general terms, the positions that are subject to testing under the authority of each agency.

In a 1988 Gallup poll, 23 percent of employers with drug testing programs reported that they tested all employees. Of those companies that tested only some employees, 43 percent tested "for cause," 10 percent tested those in security-sensitive positions, 20 percent tested drivers, and 12 percent performed random testing. A total of 86 percent of respondents tested applicants, and 81 percent of these tested all applicants.[46]

Communication of the Policy

Effective communication of the policy can increase employee awareness of substance abuse problems, can enhance employee acceptance, and can help the program withstand legal scrutiny. The policy should explicitly state the separate responsibilities of the employer and the employees. An employee's right to refuse testing and the consequences of such refusal should also be stated. The circumstances, if any, in which retests may occur should be noted. The employer should give a copy of the policy to each employee, and should request written acknowledgment of having received and understood the policy. Employees should also receive periodic reminders of the policy, for example, in company newsletters.

Types of Tests

Preemployment/Preappointment
Testing can occur before a formal offer is extended, or after, with the offer contingent on successful completion of a drug test. Testing of all applicants, even if not in safety- or security-sensitive positions, avoids the appearance of discrimination. Such testing has been almost universally upheld by the courts.

Reasonable Cause
Reasonable cause, also known as *reasonable suspicion* and *for cause*, testing refers to testing of an individual who is reasonably suspected of drug use based on physical, behavioral, or performance indicators. *Probable cause* implies a higher degree of individualized suspicion than reasonable cause. Among other things, reasonable cause testing may be based on:[50]

1. Observable phenomena, such as direct observation of drug use and/or the physical symptoms of being under the influence of a drug;
2. A pattern of abnormal conduct or erratic behavior;
3. Arrest or conviction for a drug-related offense, or being the target of a criminal investigation into a drug-related offense*;

---

*Courts have ruled that testing because of an arrest, conviction, or criminal investigation for a drug-related crime is valid only if the offense is "recent."

4. Information provided by reliable and credible sources or independently corroborated;
5. Evidence of tampering with a previous drug test specimen.

Some of these terms—for example, "erratic behavior"—are unclear and subjective. The ability of supervisors to monitor employees for signs of drug abuse has also been questioned, thus pointing to the need to clearly define what constitutes a performance problem for reasonable cause testing. Reasonable cause does not require certainty, but it does require more than a hunch. Reasonable cause testing has generally passed legal scrutiny when there is clear evidence of on-duty use or impairment.

In companies that perform reasonable cause testing, supervisors should be trained to identify behavior and objective job performance decrements that could be drug-related. Laypeople can miss symptoms of drug use. For alcohol, reasonable cause tests have generally been based on poor job performance indicators rather than behavioral observations. It is best if supervisors base their decision to test on job performance rather than attempting to diagnose drug abuse. Testing should be made contingent upon concurrence from higher-level management that reasonable suspicion exists. Some employer-union agreements also require a union representative's involvement. Each supervisor should measure and document job performance regularly, and should have appropriate records to back up any performance-related decisions to test.

Postrehabilitation/Extended Absence
An employer who requires counseling or rehabilitation treatment as a condition of continued employment may wish to monitor the individual both during and after rehabilitation. DOT authorizes postrehabilitation, unannounced testing for a period of up to 5 years. Some employers also test individuals who return after a prolonged absence from work.

Post-accident/Post-incident
The employer's policy should define those accidents or safety violations that trigger a drug test. Each DOT modal regulation defines the accidents, incidents, and/or safety violations that trigger a DOT drug test. Involved victims are typically also tested.

Periodic
Many employers require routine periodic physical examinations as a condition of continuing employment. Employers may include drug testing as part of such examinations. The use of a drug screen as part of a voluntary examination program may, however, dissuade employees from participation,

with consequent loss to the nonparticipants of the benefits of the examination. Potential health benefits should be weighed against the need for drug testing before a decision is reached on this matter.[56]

Random

Announced drug testing has limited value in deterring or identifying drug abuse. Given ample announcement time, a drug abuser may defeat a test by any number of means. Unannounced, random testing, on the other hand, has proven a strong deterrent to drug abuse. Well-designed random testing programs are appropriate and legally defensible in certain circumstances, for example, where legally mandated, or where there is a significant safety or security need.

Much of the controversy surrounding random testing is based on the common, but incorrect, perception that *random* means "arbitrary." In fact, random testing eliminates much of the potential for discretion on the part of management. The purpose of random testing is to protect employees from being singled out for testing for arbitrary or discriminatory purposes, as well as assuring that all covered employees are equally at risk for being tested. The impartial nature of random testing can be reinforced by having people without supervisory or hiring/firing responsibility generate random selections.

DOT's random testing requirements authorize a 50 percent annual testing rate. NRC's random requirements, and the proposed DOE random requirements, require testing at a 100 percent annual rate, resulting in about two-thirds of the workers being tested during any given year. Among federal agencies that conducted random testing in 1990, annual rates ranged from 4 percent to 200 percent, with most being between 10 percent and 100 percent.[57]

An employer with 100 covered employees and a 50 percent random testing rate would make 50 random selections during the year to achieve this rate. Although statistically improbable, one person among the 100 in this example could theoretically be selected for testing on 2, 5, 10, or even all 50 occasions. Random selections can be better spread out by use of a "two-drum" system: All covered employees are initially in both drums. Selections are made from both drums. Names drawn from the first drum are returned to that drum, so that each person remains in the testing pool. Names drawn from the second drum are not returned to that drum; instead, the second drum is gradually depleted. The first drum helps ensure that each person remains subject to testing; the second drum helps ensure that each person is eventually tested.

Other common methods of random selections use a computerized random number generator, or a table of random numbers, or a single drum from which names are selected. Random selections should be reasonably spread throughout the year, for example, selections should be made on a quarterly or monthly basis.

Summary

Reasonable cause, post-accident, and postrehabilitation testing are triggered by the employee's acts. Random and periodic testing are triggered independent of the employee. There have been far fewer legal challenges to reasonable cause, post-accident, and postrehabilitation testing than to other types of testing.

Court rulings have generally upheld all of the aforementioned types of drug testing for government employees in safety- or security-sensitive positions. For employees in nonsensitive positions, preemployment testing has been upheld, reasonable cause testing has been upheld if based on job-related impairment, and post-accident testing has been upheld if based on individualized suspicion. Random testing has met with mixed results in legal challenges. For nonregulated, private-sector employers, all of the above types of testing are probably acceptable when uniformly and appropriately applied, except for random testing of non-safety-sensitive positions.

What to Test For

Illicit substances, alcohol, and legitimately prescribed drugs—for example, barbiturates and benzodiazepines—can pose problems in the workplace. A comprehensive substance abuse policy addresses all of these. If prescription drugs are tested for, the employer should not discriminate against those who have valid medical reasons for using certain drugs. The employer's policy may dictate that while employees are using prescription medications that can impair job performance, they are not permitted to work in certain safety-sensitive positions.

The NIDA Guidelines and DOT Procedures authorize testing for five drugs and classes of drugs: cocaine, marijuana, amphetamines, opiates, and PCP. These regulations also state the cutoff levels for each drug or drug class. NRC and certain DOT agencies also authorize alcohol testing. Employers regulated by these federal agencies may test for other drugs and drug classes, and may test at other cutoffs, but can do so only outside the authority of the federal regulations, which means there must be a separate drug testing program, with separate specimen collections, for the nonregulated tests.

The employer may tailor the program to those drugs that are relevant to the work force. Local law enforcement and public health officials can provide information on drugs that are currently the greatest problem in the area. The drugs that are tested for, and their respective initial and confirmation cutoff levels, should be indicated in the employer's policy.

On-Site Versus Off-Site Testing

On-site testing provides results quickly and at a potentially reduced cost. The NRC regulations permit on-site initial testing, but require that confirmatory testing be done at NIDA-certified laboratories. Other federal agency regulations require that both initial and confirmatory analyses be conducted only at NIDA-certified laboratories. On-site testing is done most often at nuclear power plants, military facilities, federal courts, corrections institutions, treatment programs, and at some large companies. Positive on-site initial tests, in any setting, should be followed up by confirmatory tests at an outside laboratory. An employer may temporarily remove an employee from a safety-sensitive position pending the results of confirmatory analysis. In doing so, the confidentiality of the initial results must be maintained.

Procedures for Challenging Results

Donors should have access to all records regarding their test results, and should be told of any opportunities to explain or challenge a positive result. Some employers allow donors to have their specimens reanalyzed at another laboratory, but may insist on certain criteria—for example, NIDA certification—of that laboratory. Some employers conduct only initial—for example, immunoassay—tests on applicant specimens; if the initial result is positive, the applicant has the option of requesting and paying for confirmatory testing. If the confirmation test result is negative, the test's outcome is considered "negative" and the applicant is reimbursed for the cost of the confirmation test. This policy allows the applicant the protection of confirmatory analysis without burdening the employer with confirmatory analyses costs on every specimen that screens positive. This approach is common in the corrections sector, but is unacceptable in federally mandated testing.

Each donor should have an opportunity to explain legitimate reasons for a laboratory-positive drug test result. The federal drug testing regulations require that a physician, functioning in the Medical Review Officer (MRO) role, receive all results and report laboratory-positive results to the employer only after reviewing them for potential legitimate explanations.

Safe Harbor for Voluntary Referrals

E.O. 12564 permits each federal agency to exempt from disciplinary action—that is, create a "safe harbor" for—employees of that agency who voluntarily identify themselves as drug abusers, obtain counseling or rehabilitation through an EAP, and thereafter refrain from drug abuse.[8,50] Since the key to this provision's effectiveness is an employee's willingness to admit his or her problem, this provision is unavailable to an employee who requests safe harbor after being asked to undergo a drug test or after testing positive.

Consequences for Positive Test Results

The primary emphasis of workplace drug testing programs should be rehabilitative rather than punitive for employees who test positive. The potential of disciplinary actions can nevertheless be a strong deterrent to drug abuse. Disciplinary actions should be taken only for confirmed positive results that have no legitimate explanation. By comparison, in the criminal justice sector, where the standard of proof (e.g., "some evidence" or "the preponderance of evidence") is less than that in workplace drug testing (e.g., "beyond a reasonable doubt"), screening test results have been upheld in court.

The federal regulations do not limit employers' discretion in responding to positive drug tests. Certain federal agencies prohibit employees from specific job functions if they have tested positive, for example, from flying an airplane or driving a truck. Beyond this, the employer decides the consequences for persons who test positive.

Few employers hire applicants who test positive. Some employers will, under certain circumstances—for example, after a waiting period, or if the applicant presents evidence of successful substance abuse treatment—allow an applicant to reapply after having tested positive. Of 1,633 firms surveyed in 1991, 96 percent did not hire applicants who tested positive.[47] One-third of the firms in this survey dismissed employees who tested positive, though in many firms dismissal was a policy of last resort for repeated abuse. A 1988 Gallup poll indicated that 17 percent of employers surveyed dismissed employees who tested positive, and 41 percent referred such employees to treatment and/or counseling.[46] Other employer options include written reprimand, "on-leave" status, probation, and suspension. Although an employer may wish to invoke the "at-will" employment doctrine (whereby an employer may terminate an employee for good reason, poor reason, or no reason), the employer may still be subject to limitations imposed by public policy exceptions to this doctrine. Legal counsel should be consulted on this matter.

Federal drug testing regulations require covered employers to provide access to treatment and rehabilitation programs, but do not require that the employer pay for such programs. Counseling and treatment programs may prevent the loss of skilled and experienced employees.

A refusal to undergo testing is usually treated the same as a positive test result. Employees should understand their right to refuse testing and the consequences of such refusals.

Confidentiality and Privacy Protections

Specimen collection in a medical setting can reduce privacy concerns. Specimens should not be collected under direct observation unless there is reason to believe an employee may tamper with or substitute the sample. These reasons should be clearly spelled out. For example, the NIDA Guidelines indicate that a urine temperature outside the range of 90.5°–99.8°F is reason to suspect that the donor may have altered or substituted the specimen.

The confidentiality of individual drug test results is protected by certain state laws and, in federally mandated testing, under the applicable federal regulations. Information regarding individual drug test results should be limited to those with a need to know, for example, the medical director, the employee's supervisor, and the head of personnel. Medical information ancillary to the actual test result—for example, medications taken by the employee—should be available only to medical personnel. Individual drug test records should be released only in response to a written authorization from the donor or the donor's representative, a subpoena for test results, or as otherwise required by law. The employer's policy should also address the handling of job references for employees who have had positive drug test results.

Employer Record Keeping

Records of individual drug tests of the program as a whole should be carefully maintained. Separating individual drug test records from other employee medical records helps preserve the distinction between drug test results and medical examinations. Records of all donors—applicants and employees, positives and negatives—should be retained because of the potential for legal claims, including discrimination claims.

Each individual test record should include a copy of the laboratory test report, a copy of the chain of custody form, all consent and release forms—for example, consent to testing, consent to releasing the results, and consent to reinstatement conditions—and any information that is obtained regarding medications taken by the donor. The employer's MRO may maintain at least a portion of these records for the employer. In federally mandated testing, results of positive results must be maintained for 2–5 years, and results of negative results must be maintained for at least 1 year.

NIDA-certified laboratories provide monthly summary reports of all test data conducted under specific accounts. These summaries include the number of specimens tested and the results of initial and confirmatory analyses for each drug or drug class. The individuals tested are not identified on these reports.

In programs with on-site testing, the employer should maintain the following records: certifications of instrument operators, test and calibration data, quality control data, performance testing data, and maintenance and service records.

EVALUATION OF DRUG TESTING PROGRAMS

Employers in a wide variety of industries have initiated drug testing programs for a variety of reasons. In the early 1980s, employers started testing in response to several well-publicized drug-related accidents, because of a desire to have a healthy and productive work force, and as a response to the growing recognition that drug abuse was a significant problem in the workplace. More recently, defensive testing has become more common—that is, employers would conduct drug testing to avoid hiring drug abusers who were being rejected by other companies that conducted drug testing. As described earlier in this chapter, in the late 1980s, federal regulations mandated testing for large segments of the American (and some foreign) work force. Regardless of the reason or motivation for implementing drug testing, the employer should regularly evaluate the program.

Costs

In the ideal situation, an employer may have compiled statistics on the contribution that alcohol and drug use by employees made toward "the bottom line." Some employers have examined the relationship between substance abuse and accidents, workers' compensation claims, absenteeism, insurance claims, productivity, and other quantifiable factors. Such studies have provided a baseline against which to measure the economic impact of drug testing. The savings gained through lowered costs or lost time can then be compared to the costs of the program.

Drug testing programs have a number of costs, some more obvious than others. Most larger employers contract directly with a laboratory, each of which offers a variety of testing panels and protocols and different prices, often dependent on the volume of specimens that will be tested. Some collection sites, laboratories, and MROs offer discounted package prices that cover several testing services. These *turn key* and *consortium* packages may offer small companies the advantage of volume discounts through the grouping of several employers that form the consortium.

Laboratory Analysis
The U.S. Government Accounting Office (GAO) assessed costs of several federal agency drug testing programs in 1990.[57,58] GAO found significant variations in the drug testing costs across 18 agencies, with combined screening and confirmation test prices ranging from $8.90 to $25.00, $8.00 to $47.49 for screening alone, and $14.00 to $66.00 for confirmation alone.

Medical Review Officer Services
In the GAO study, MRO services ranged from $50 to $200 per hour. In a 1990 survey of several hundred MROs,[59] typical charges were $5 to $15 for administrative review of laboratory negatives; $35 to $75 for review of laboratory positives; $15 to $25 combined fee, for any result.

Collection
In the 1990 survey of MROs, most of whom also provided collection services, typical collection and shipment charges were $10 to $20. When the collection site is not readily serviced by the laboratory's courier network, additional costs are incurred in shipping the specimen—usually by overnight courier service—to the laboratory. Some companies collect specimens in-house—for example, at the medical department—which is usually cheaper than collections done elsewhere. Collection services appear to get less attention and money than other program components probably because collections are perceived as less technical. However, more errors and financial losses occur at this point than in the entire process. Specimens that are invalidated—that is, are unusable—are usually faulted for collection errors. The donor is usually called back for another collection. Properly performed collections avoid the lost time and excess charges that may occur from improper collections. For these reasons, the collection process is a good place to look for cost savings.

Employee Assistance Program
EAPs assist employees with drug abuse or other problems by means of counseling, treatment, or referral to treatment. EAPs usually cost more than drug testing programs, but receive less scrutiny. There are no certification programs for EAPs, so it is especially important for the employer to periodically evaluate its EAP services. Various techniques are available, such as surveys of employee awareness of what services are available, EAP client satisfaction, and counseling or treatment success. The EAP should

provide the employer with periodic reports on numbers of referrals, statistical description of problems encountered, and other ongoing activities.

Treatment
Substance abuse treatment appears to receive the least scrutiny from employers. Treatment services are almost always contracted out. The responsibility for oversight and monitoring of the treatment services is often left to the insurer. In any event, poor treatment will not be effective enough to pay for itself. More attention should be paid to the selection of quality treatment providers, who should be periodically evaluated to gauge effectiveness.

Impact on Employee Health and Safety

The two most frequently used justifications for implementing drug testing programs are to improve workplace safety and to maintain a healthy work force. Too often these worthy goals are lost in labor disputes that focus on traditional "we versus they" issues. Employees and their labor representatives may perceive that drug testing is a device the employer uses to get rid of employees. An important technique available to employers is to thoroughly assess the actual health and safety issues affecting their work force. Information and statistics are usually available for a period of time prior to implementing any drug testing program. The number of accidents, workers' compensation costs, medical costs, and sick leave benefits can be measured. These same data should be collected on an ongoing basis, providing a direct measure of achieving these program goals. If employees are given objective information about a reduction in accidents and improvement in the general welfare of the work force, they are more likely to support the program.

Impact on Employee Morale

The most successful drug programs are initiated with a carefully planned and deliberate approach that includes extensive employee communication. This approach makes employees aware of the reasons their employer conducts drug testing, and the consequences and available options when an employee tests positive. When programs are implemented in haste, employees may feel that drug testing has been "rammed down their throats." Concerns about constitutional rights and accuracy of testing are only heightened by a poorly communicated program, regardless of how positive the goals are. Employers can do many things to help improve employee relations and morale.

The three C's of labor relations—communicate, communicate, communicate—are important. Many employers have done much to alleviate employee concerns by developing or purchasing videos that provide accurate and objective information about the company's program. Sometimes it is advisable to have representatives from the various program components—for example, the laboratory, EAP, quality assurance program, and medical staff—participate in employee seminars. These can be held periodically or videotaped for use at employee meetings.

Employers should make available pamphlets and other written information about the EAP and other services. Most EAPs will provide such literature as part of their services. Under the federal regulations, the EAP must provide a minimum of training of supervisors who may be involved in identifying impaired employees. An even more effective approach is to provide all employees with similar training.

One of the most important steps an employer can make to ensure the success of a drug testing program and gain employee support is to thoroughly evaluate the need for it. The policies and procedures must be carefully developed to meet the need and fit the specific work environment. There are no "canned" programs. The programs that work best are tailored for each employer. The program should be developed with employee input. Then it must be communicated well before implementation. Drug and alcohol programs are successful when everyone works together for a common cause.

SUMMARY

Drug testing has become a valuable part of employers' efforts to create drug-free environments in their workplaces. Employers in the private and public sectors have found that carefully designed and monitored drug testing programs can save money and employees lives.

References
1. Hansen, H. J., S. P. Caudill, and J. Boone. Crisis in drug testing: Results of CDC blind study. *JAMA* 1985;253:2,382-2,387.
2. 49 CFR § 391.41 through 391.49.
3. *EEI Guide to Effectiveness. Drug and Alcohol/Fitness for Duty Policy Development.* Washington, D.C.: Edison Electric Institute; 1985.
4. Burt, M. R., M. M. Biegel, Y. Carnes, and E. C. Farley. *Worldwide Highlights From the Worldwide Survey of Non-Medical Drug Use and Alcohol Use Among Military Personnel.* Bethesda, Md.: Burt Associates; 1980.
5. Mulloy P. J. Winning the war on drugs in the military. In: R. H. Coombs and L. J. West (eds.), *Drug Testing: Issues and Options.* New York: Oxford University Press; 1991:92-112.
6. *Relative Effectiveness and Impact of Navy Drug Control Initiatives.* U.S. Navy

Contract Report N00600-82-D-2956. Washington, D.C.: Booz-Allen & Hamilton, Inc.; 1984.
7. 42 CFR § 74.
8. Reagan, R. Executive Order 12564: Drug-free federal workplace. *Fed. Reg.* 1986;51(Sept. 27):32,889-32,893.
9. Supplemental Appropriations Act of 1987. Section 503, pub. L. 100-71, 101st Congress, July 11, 1987.
10. Scientific and technical guidelines for federal workplace drug testing programs; standards for certification of laboratories engaged in urine drug testing for federal agencies: Notice of proposed guidelines. *Fed. Reg.* 1987;52(Aug. 14):30,638-30,651.
11. Department of Health and Human Services (DHHS). Mandatory guidelines for federal workplace drug testing programs. *Fed. Reg.* 1988;53(April 11): 11,970-11,989.
12. Finkle, B. S., R. V. Blanke, and J. M. Walsh (eds.). *Technical, Scientific and Procedural Issues of Employee Drug Testing.* Rockville, Md.: National Institute on Drug Abuse; 1990:21-24.
13. U.S. Department of Transportation. Procedures for transportation workplace drug testing programs: Interim final rule. *Fed. Reg.* 1988;53(Nov. 21):47,002-47,021.
14. U.S. Department of Transportation. Procedures for transportation workplace drug testing programs: Final rule. *Fed. Reg.* 1989;54(Dec. 1):49,854-49,884.
15. U.S. Department of Defense. Drug-free work force. *Fed. Reg.* 1988;53(Sept. 28):37,763-37,765.
16. U.S. Department of Defense. Drug-free work force. *Fed. Reg.* 1991;56(Nov. 27):60,066—60,073.
17. Nuclear Regulatory Commission. Fitness-for-duty programs. *Fed. Reg.* 1989; 54(June 7):24,468-24,508.
18. U.S. Department of Energy. Workplace substance abuse programs at DOE facilities: Proposed rule. *Fed. Reg.* 1991;56(July 3):30,644-30,652.
19. U.S. Department of Energy. Personnel security assurance program; substance abuse testing procedures: Notice of proposed rulemaking. *Fed. Reg.* 1991;56(March 8):10,075-10,079.
20. Baer, D. M., R. E. Belsey, and M. R. Skeels. A survey of state regulation of testing for drugs of abuse outside of licensed (accredited) clinical laboratories. *Am. J. Public Health* 1990;80:713-715.
21. *A Guide to State Drug Testing Laws & Legislation.* Washington, D. C.: Institute for a Drug-Free Workplace; 1991.
22. Drug Testing Quality Act of 1991. H.R. 33, 102nd Congress, 1st Sess., sec. 1, 1991.
23. Quality Assurance in the Private Sector Drug Testing Act of 1991. S. 2008, 102nd Congress, 1st Sess., sec. 1, 1991.
24. Hanson, D. Drug abuse testing programs gaining acceptance in the workplace. *Chemical and Engineering News* 1986;64:7-14.
25. Willette, R. E. Drug testing programs. In R. L. Hawks and C. N. Chiang (eds.),

Urine Testing for Drugs of Abuse. Rockville, Md.: National Institute on Drug Abuse; 1986:5-12.
26. Osborn, C. E., and J. J. Sokolov. Drug trends in a nuclear reactor company: Cumulative data from an ongoing testing program. In S. W. Gust and J. M. Walsh, (eds.), *Drugs in the Workplace: Research and Evaluation Data*. Rockville, Md.: National Institute on Drug Abuse; 1989:69-80.
27. Crouch, D. J., D. O. Webb, L. V. Peterson, P. F. Buller, and D.E. Rollins. A critical evaluation of the Utah Power and Light Co substance abuse management program: Absenteeism, accidents, and costs. In S. W. Gust and J. M. Walsh, (eds.), *Drugs in the Workplace: Research and Evaluation Data*. Rockville, Md.: National Institute on Drug Abuse; 1989:97-108.
28. Taggart, R. W. Results of the drug testing program at Southern Pacific Railroad. In S. W. Gust and J. M. Walsh, (eds.). *Drugs in the Workplace: Research and Evaluation Data*. Rockville, Md.: National Institute on Drug Abuse; 1989:97-108.
29. *Summary of Findings from the 1990 National Household Survey on Drug Abuse*. Rockville, Md.: National Institute on Drug Abuse; 1989.
30. *National Treasury Union Employees' Union v. Von Raab*, 109 S. Ct. 1384 (1989).
31. Lund, A. K., D. F. Preusser, R. D. Blomberg, and A. F. Williams. Drug use by tractor trailer drivers. *J Forensic Sci*. 1988;33:648-661.
32. *Skinner v. Railway Labor Executives' Association*, 109 S. Ct. 1402 (1989).
33. Bureau of Labor Statistics. *Survey of Employer Anti-Drug Programs*. Washington, D.C.: U.S. Department of Labor, Report 760; January 1989.
34. SmithKline Beecham Clinical Laboratories, King of Prussia, Penna.; 1992.
35. *Seventh Special Report to Congress on Alcohol and Health*. Rockville, Md.: National Institute on Drug Abuse; 1990.
36. Gust, S. W., and J. M. Walsh, (eds.). *Drugs in the Workplace: Research and Evaluation Data*. Rockville, Md.: National Institute on Drug Abuse; 1989.
37. Chait, L. D., M. W. Fischman, and C. R. Schuster. Hangover effects the morning after marijuana smoking. *Drug and Alcohol Dependence* 1985;15:229-238.
38. Yesavage, J. A., V. O. Leirer, M. Denari, and L. E. Hollister. Carry-over effects of marijuana intoxication on aircraft pilot performance: A preliminary report. *Am. J. Psych*. 1985;142:1,325-1,329.
39. Leirer, V. O., J. Yesavage, and D. Morrow. Marijuana carry-over effects on aircraft pilot performance. *Avia. Space Environ. Med.* 1991;62:221-227.
40. Parish, D. C. Relation of the pre-employment drug testing result to employment status. *J. Gen. Intern. Med*. 1989;4:44-47.
41. Normand, J. Relationship between drug test results and job performance indicators. Presented at the 97th American Psychological Association meeting, New Orleans, La.; 1989.
42. Normand, J and S. Salyards. An empirical evaluation of preemployment drug testing in the United States Postal Service: Interim report of findings. In S.W. Gust and J.M. Walsh (eds.). *Drugs in the Workplace: Research and Evaluation Data*. Rockville, Md.: National Institute on Drug Abuse; 1989:111-138.
43. Zwerling, C., J. Ryan, and E. J. Orav. The efficacy of preemployment drug

screening for marijuana and cocaine in predicting employment outcome. *JAMA* 1990;264:2,639–2,643.
44. Wish, E. D. Preemployment drug screening. *JAMA* 1990;264:2,676–2,677.
45. Blank, D. L., and J. W. Fenton. Early employment testing for marijuana: Demographic and employee retention patterns. In S. W. Gust and J. M. Walsh (eds.), *Drugs in the Workplace: Research and Evaluation Data.* Rockville, Md.: National Institute on Drug Abuse; 1989:139–150.
46. *Drug Testing at Work: A Survey of American Companies.* Gallup Organization; 1988.
47. *1991 AMA Survey: Workplace Drug Testing and Drug Abuse Policies.* New York: American Management Association; 1991.
48. *Wall Street Journal,* March 3, 1991.
49. Federal personnel manual letter 792-19. *Fed. Reg.* 1989;54(Nov. 13):47,324.
50. National Institute on Drug Abuse. *Model Plan for a Comprehensive Drug-Free Workplace Program.* Rockville, Md.: U.S. DHHS publication no. (ADM)89-1635; 1989.
51. Kasler, P. Legal issues in drug testing: A review of recent cases. *Syva Monitor* 1989;7(1).
52. Shults, T. (ed.). *Employee Testing and the Law.* Chapel Hill, N.C.: Vanguard Information Publications; 1990.
53. Hoyt, D., R. Finnigan, T. Nee, T. Shults, and T. Butler. Drug testing in the workplace—Are methods legally defensible. *JAMA* 1987;258:504–509.
54. Hatch, O., and V. Schachter. Drugs of abuse testing: A legislative and legal update. Presented at the American Association of Clinical Chemistry meeting, San Francisco, Calif., July 25, 1990.
55. *Testing for Substance Abuse: A Policy and Procedure Guide for Industry.* Palo Alto, Calif.: Syva Co.; 1986.
56. American College of Occupational Medicine. Drug screening in the workplace: Ethical guidelines. *J. Occup. Med.* 1991;33:651–652.
57. *Employee Drug Testing Status of Federal Agencies' Programs.* Washington, D.C.: U.S. Government Accounting Office publication no. GAO/GGD-91-70;1991.
58. *Employee Drug Testing. A Single Agency Is Needed to Manage Federal Employee Drug Testing.* Washington, D.C.: U.S. Government Accounting Office publication no. GAO/GGD-91-25;1991.
59. *Medical Review Officer Information Handbook.* Arlington Heights, Ill.: American College of Occupational Medicine; 1991.

3
Drug Testing Collection Procedures

Robert Swotinsky, M.D., M.P.H, and
Janet J. Beatey, M.T. (A.S.C.P.)

Specimen collection is the most vulnerable part of workplace drug testing procedures. The validity of a drug test result depends on protecting both the integrity and proper identity of the specimen. Well-designed collection procedures, such as those described in the federal drug testing regulations, accomplish this while minimizing any intrusion on the donor's privacy.

The first set of federal drug testing procedures appeared in the Department of Health and Human Services (DHHS) "Mandatory Guidelines for Federal Workplace Drug Testing Programs"[1] (also known as the NIDA Guidelines). These were precedent-setting procedures that were required of federal agency programs and were voluntarily adopted by many other employers. The Nuclear Regulatory Commission (NRC) regulations[2] included procedures similar to, but more permissive than, those of the NIDA Guidelines. Subsequently, the U.S. Department of Transportation (DOT) published more elaborate and rigorous drug testing procedures for the transportation industries.[3] The DOT Procedures have since replaced the NIDA Guidelines as the "gold standard." This chapter presents collection procedures consistent with the federal regulations as of 1991.

Drug test collection procedures are designed to meet forensic standards—that is, to be legally defensible. For the results of a particular specimen to withstand legal scrutiny, it is necessary to demonstrate:

1. No adulteration or tampering has taken place;
2. Documentation is listed of all persons who handled the specimen;
3. No unauthorized access to the specimen was possible;
4. The specimen was handled in a secure manner;

5. The specimen belongs to the individual whose identification is printed on the specimen container's label.

Handling errors during collections are the most common and most successfully challenged deficiencies in urine drug testing.[4] There is a greater potential for human error in this process than in the actual laboratory analysis of the specimen. The collector therefore has a pivotal role in each drug test.

The collector typically works for the laboratory, the employer, or the employer's medical provider. The employee's direct supervisor should not serve as the collector unless there is no feasible alternative. Specimens collected by law enforcement officials and accident investigators are usually inconsistent with the employer's policy and procedures, and are therefore often unacceptable.

Forensic collection and testing procedures have recordkeeping requirements that are more rigorous than those of diagnostic or screening test procedures. Therefore, individuals should be trained before they start collecting specimens for forensic drug testing. DOT and several commercial organizations have offered training sessions for collectors. DOT requires that nonmedical people who collect DOT drug test specimens first receive training and demonstrate their competence in this function. Any individual who collects a DOT drug test specimen must first receive written collection instructions, preferably in the form of a checklist, for example, as presented in this chapter.

Collection procedures involve the most sensitive and difficult aspects of drug testing. It is here that the competing values of privacy and confidentiality are most in conflict with the need to ensure integrity and proper identity. The collector should have a professional attitude and sensitivity to the concerns of the donor, because without the collector's technical and professional expertise, the entire program is subject to criticism.

COLLECTION SITE PREPARATION

Selection

A collection site may be located at a medical facility, a mobile trailer, an employer's office, or anywhere else that has:

An enclosure where privacy for urination is possible;
A toilet for completion of urination (unless a single-use collection container is used with sufficient capacity to contain the void);
A water source for washing hands, preferably outside of the toilet stall;
A surface to write on, for completing the chain of custody form;
A secure place to leave personal items such as purses and briefcases;
Restricted access so that the site is secure during collection.

Security

The collection site must be prepared so that the possibility of a sample being diluted or adulterated is minimized. Unauthorized people must be restricted from the site during collection or storage of specimens.* Alternate entrances to the collection site—for example, back and side doors—should be blocked. Precautions must be taken to prevent the donor's access to water and other potential diluents and adulterants. The enclosure should be regularly inspected for containers of urine or other liquid that might be hidden away for use as an adulterant.

Cleansers and disinfectants should be removed from the enclosure. A bluing agent—for example, blue toilet cleanser, ink, or food coloring—must be added to the toilet water if possible. If a bluing agent dispenser has been placed in the toilet tank reservoir, the toilet should be flushed before collection to add bluing agent to the water in the bowl. Water sources, other than the toilet, must be removed, sealed, or otherwise made inaccessible to the donor within the enclosure. Faucet handles can be turned off and sealed with tamperproof tape. Alternatively, an outside valve can be shut to stop water flow to faucets in the enclosure. If the toilet does not have a bowl or tank water supply, water to the toilet should be shut off during the collection and turned back on thereafter.

Supplies

Disposable—for example, latex—gloves are recommended, but not required, for specimen handling.[5] The Centers for Disease Control recommend glove use for urine collection if the urine contains visible blood.[6]

The collector needs the following supplies:

1. Specimen bottles. The donor may void directly into the specimen bottle. If the bottle's opening is narrow, making urination directly into the bottle difficult, the collection site should also have wide-mouthed collection containers into which the donor can urinate. The collector then pours the urine from the collection container into the specimen bottle. Each container must hold at least 60 ml, the minimum volume required by federal regulations. DOT requires that all collection containers be single-use and be securely wrapped or sealed until opened in the donor's presence. Specimen bottles

---

*DOT permits a union or legal representative to be present during the collection if the employer's policy permits such practice, and if the presence of the representative does not disrupt or interfere with the collection process. The representative must remain outside the stall or bathroom while the donor is providing the specimen. DOT does not permit more than one donor to provide a specimen at the same time in a multiuse bathroom.

must resist punctures; the donor could otherwise place a pinhole through the bottle, allowing urine to leak out during shipping.

2. A thermometer (or other temperature-sensing device) for measuring urine temperature. This thermometer must provide an immediate reading. A disposable, liquid crystal thermometer may be attached to the side of the collection container—preferably the outside—and poses little risk of contaminating the specimen. A digital thermometer requires a disposable tip to prevent contamination. Standard mercury thermometers are unsuitable because they react relatively slowly to temperature changes.

3. A thermometer for measuring the donor's oral temperature.

4. Tamperproof seals for specimen bottles and shipping containers. A tamperproof seal is a tape strip that breaks easily under tension but does not stretch or peel. A seal can also serve as the label on which to identify a specimen's collection number, the donor's initials, and the date. Alternatively, a separate label can be attached to the container for this purpose.

5. A shipping container in which to send the specimen bottle(s) and associated chain of custody forms to the laboratory. The container must be designed to minimize the possibility of damage during shipment, and must allow for secure sealing.

6. Bluing agent to add to the toilet bowl or tank to discourage dilution of the specimen.

7. A chain of custody form (also known as a custody and control form, or COC form). This is used to maintain control and accountability of a urine specimen from the collection site to the laboratory.

8. The collection site may need a book in which to record identifying data on each specimen. The NIDA Guidelines and NRC regulations require sequential entry of collection data into a permanently bound record book. Figure 3-1 is a form that can be bound and used for this purpose. Many collections are conducted with chain of custody forms—such as the DOT-mandated form—on which these data are also entered. When the chain of custody form captures these data, a bound record book is unnecessary.[7]

Chain of Custody Forms

Each chain of custody form includes these components:

1. A chain of custody block. The name of each person who handles or transfers the specimen, the reason for such handling or transfer, and the

FIGURE 3-1. Drug test collection log.

I agree to submit to a urine drug screen. I certify that the urine specimen identified on this form is my own; that it is fresh and has not been adulterated in any manner. The identification information provided on this line and on the collection bottle is correct. I understand that this urine specimen will be analyzed for drugs. I further consent and agree that the results of this analysis will be furnished by the laboratory or medical facility solely to me, my employer, my employer's Medical Review Officer, and the appropriate state agencies where required by applicable law and regulations. The results will not be released to additional parties without my written authorization. I agree to hold the laboratory that analyzes my specimen(s) and the collection site harmless from any action that may arise out of my testing as defined herein other than from their negligence. My signature below ("Donor's Signature") acknowledges that I have read and understand the foregoing.

DATE	DONOR'S NAME	DONOR'S SOCIAL SECURITY NUMBER	SPECIMEN NUMBER	TYPE OF TEST	TEMP	REMARKS	COLLECTOR'S SIGNATURE	DONOR'S SIGNATURE	CLIENT

PRE = PRE-EMPLOYMENT
RC = REASONABLE COURSE
RAN = RANDOM
RTD = RETURN TO DUTY
PA = POST ACCIDENT
OT = OTHER OR NOT SPECIFIC
PM = PERIODIC MEDICAL

PURPOSE OF CHANGE	RELEASED BY	RECEIVED BY	DATE
Provide Specimen for Testing	- DONOR -	*(Collector)*	*(Date)*
(Shipment to Lab)	*(Collector)*	*(Courier)*	*(Date)*

FIGURE 3-2. DOT chain of custody block. (Entries in italics are completed at the collection site.)

date of each person-to-person transfer are recorded in this section. Figure 3-2 shows the chain of custody block used in DOT-mandated testing.

2. The specimen identification code. This is a unique identifier that is also placed on the specimen bottle. It must be preprinted or typed, not handwritten, on DOT chain of custody forms.

3. The donor's identification. This can be the donor's name, employee I.D number, social security number, or some other identifier. DOT chain of custody forms have multiple copies; the copies sent to the laboratory do not have the donor's name, so the laboratory identifies the specimen by social security number.

Most laboratories provide chain of custody forms for their clients. Chain of custody forms printed for specific employers usually display employer-specific information, for example, billing codes, the employer's name, and the Medical Review Officer's (MRO) name. Chain of custody forms usually have certification statements that, when signed by the donor, collector, or both, attest to the specimen's integrity.

DOT Chain of Custody Forms

DOT imposes detailed requirements on the design and contents of chain of custody forms. A form that meets DOT's requirements is illustrated at the end of the DOT Procedures (see Appendix C). DOT requires a seven-copy carbonless form; Copy 7 is used only when split samples are collected. (Split-sample collection is discussed later in this chapter.) The copies are distributed by the collection site as follows:

Copies 1 and 2: Accompany the specimen to the laboratory. (The laboratory subsequently sends copy 2 to the MRO.)
Copy 3: Is sent to the MRO.
Copy 4: Is given to the donor at the collection site.
Copy 5: Is retained by the collection site.

Copy 6: Is sent to the employer.
Copy 7: Accompanies the split specimen to the laboratory.

Non-DOT Chain of Custody Forms

There is no standard design or required content of chain of custody forms in non-DOT drug testing. Non-DOT forms usually have fewer copies and hold less information. Some non-DOT programs use the chain of custody format developed by DOT; however, DOT has discouraged this.

Consent and Release Forms

Before collection each donor should, in writing, consent to the test and release the results to the MRO and to the employer. This is usually done by signing a consent and release form. Some collection sites have used these forms to seek written indemnification from the donor. NRC requires the use of consent forms. DOT does not prohibit their use, but does prohibit asking the donor to "waive liability with respect to negligence on the part of any person participating in the collection, handling, or analysis of the specimen or to indemnify any person for the negligence of others.[8] A written release form should contain the following[9]:

1. The examinee's name and signature;
2. The date of the written authorization;
3. The name of the individual or organization that is authorized to release the information;
4. The name of the individual or organization that is authorized to receive the information;
5. A description of the information that is authorized to be released;
6. A description of the purpose for releasing this information.

Figure 3–3 presents an example of a release and consent form.

Medication Lists

The donor may be asked at the collection site to record medications taken before the test. In DOT-mandated testing, donors have the option of preparing such a list to serve as a "memory jogger," to help them recall the medications later should they be the cause of a laboratory-positive result. In DOT-mandated testing, the list can be placed on the donor's copy of the chain of custody form (copy 4) or a separate piece of paper for the donor's use only. This is designed to prevent the laboratory from knowing the drugs that may show up, and to avoid overreliance by the MRO on such lists.

NAME   EMPLOYEE NUMBER   WORK LOCATION

SOCIAL SECURITY NO.

HOME ADDRESS

HOME TELEPHONE NO.

In accordance with and subject to the terms and safeguards of <EMPLOYER'S> Drug Abuse Policy, I hereby acknowledge the requirements to give one or more specimens of my urine to <EMPLOYER>, its designated agents including any medical facility, laboratory or medical person, upon the request of <EMPLOYER>. This specimen shall be used solely to detect the presence of prohibited drugs.

I further consent and agree that the laboratory results of any tests performed on such specimen(s) shall be furnished by the laboratory solely to the Medical Review Officer designated in the Policy, <EMPLOYER'S> Drug Abuse Program Manager, and the appropriate federal <e.g., DOT> and state agencies where required by applicable law and regulations. I also authorize the release of this information to any physician needing it to determine my physical qualification in accordance with DOT regulations, if applicable.

I understand that <EMPLOYER'S> agents will hold the test results confidential and release them only in accordance with <EMPLOYER'S> Drug Abuse Policy. I further understand that refusal to sign this form or a positive test result may result in disciplinary action up to and including discharge.

Executed this _____ day of _____, 19____, at _____.

WITNESS SIGNATURE   SIGNATURE

FIGURE 3-3. Example of a release and consent form.

NRC, by contrast, requires that donors list all prescription and over-the-counter medicines that they can remember using within the past 30 days.

Employee Notification

Specimens should be collected soon after notification, for example, within 2 working days after notice to applicants. A post-accident DOT-mandated test must be collected as soon as possible, and no later than 32 hours after the accident. The employer is responsible for instructing employees regarding procedures for undergoing post-accident drug tests. Random tests should be collected on the same day, preferably within a few hours, of notice to selected employees.[10,11] If an employee is unavailable for a valid reason (e.g., working a different shift, on travel, or on sick leave), the employer should document the absence. In random testing, the employer generally tests at a rate equal to a specific percentage of the employee population. If a selected employee is unavailable, the employer can maintain the testing rate in one of two ways: The employer can delay notification until a time at which the employee becomes available. Alternatively, the employer can produce a surplus of random selections and choose others on that list when initially selected employees are unavailable. This latter strategy tends to exclude employees who travel frequently, thereby precluding a representative cross-section of the work force.

Donors should be notified privately of their selection for testing, should be told what type of test they have been selected for—for example, random, reasonable cause—and should be told the time they are to report, the location of the collection site, and the requirement for appropriate photo identification.

THE COLLECTION

Collection Checklists

Detailed checklists can help ensure consistent and complete collection procedures. Each donor should receive written instructions on what to expect during the collection and testing process. The collector should receive written instructions regarding his or her responsibilities. Donor and collector checklists (examples are presented in Figure 3-4 and 3-5) are a requirement in DOT testing. These checklists, particularly the collector's checklist, summarize the procedures by which specimens are collected. Checklists should be available at the collection site to both the donor and collector for reference at the time of collection.

(Text continues on page 67.)

DONOR'S CHECKLIST

Please review this description of your role in the collection of a urine specimen for drug testing. Do not hesitate to ask any questions of your employer or the collector.

- ☐ You will be asked to present a photo identification upon arrival at the collection site. If you do not have a photo ID, an employer representative will be asked to identify you.
- ☐ You can ask the collector to show his/her identification.
- ☐ You will be asked to remove any loose clothing such as a jacket, coat, or vest. Purses and/or briefcases will remain with those clothes in a secure place. You may keep your wallet. You may ask for a receipt.
- ☐ You will be asked to wash and dry your hands.
- ☐ The collector will ask you to select a sealed specimen bottle, which will be opened in your presence. If needed, you will also select a sealed collection container, which will be opened in your presence.
- ☐ You may provide the specimen alone in the privacy of an enclosure, unless the employer has specified that the specimen is to be collected under direct observation ("witnessed"), in which case the collector must watch you urinate into the container.
- ☐ You should not lose sight of the specimen until it is sealed and labeled in your presence.
- ☐ Upon receiving the specimen from you, the collector will check its volume, temperature, and appearance. If the volume is insufficient, you will be asked to drink beverages and provide another specimen. If the temperature or appearance is abnormal, you may be asked to provide a second specimen under direct observation.
- ☐ The collector will seal the bottle in your presence.
- ☐ You will be asked to initial the identification label on the specimen bottle to certify that it is your specimen.
- ☐ You will be asked to sign a statement certifying that the specimen is yours.
- ☐ You may be asked to list medications/prescriptions that you took during the past several weeks. If you are undergoing a DOT-mandated test, you can list medications only on the back of your copy of the chain of custody form.
- ☐ The results of the laboratory analysis will be forwarded to your employer's Medical Review Officer. Your employer will be notified if the laboratory results are negative. If the laboratory results are positive, the Medical Review Officer will first contact you and give you an opportunity to discuss the results and to submit information indicating the legitimate use of the drug(s) in question.

FIGURE 3–4. Donor's checklist.

COLLECTOR'S CHECKLIST

PRIOR TO COLLECTION

☐ Keep unauthorized people out of areas where urine specimens are collected or stored.

☐ Place toilet bluing agent in the toilet bowl so the reservoir of water is blue.

☐ Take steps to ensure that the collection enclosure has no other sources (e.g., shower, sink) from which water can be obtained.

☐ Remove soap and cleaning agents from the collection enclosure.

☐ Make sure that you have the necessary collection materials including specimen collection kits, shipping containers, chain of custody forms, and thermometers.

THE COLLECTION

☐ Be highly professional at all times. Avoid any conduct or remarks that might be construed as accusatorial or otherwise offensive or inappropriate.

☐ Request photo identification from the donor. If the donor has no photo ID, ask an employer representative to positively identify the donor. Document the employer representative's name, date, and time of verification. If the donor is self-employed and has no photo ID or employer representative, request a driver's license and one other signed ID, and verify that the signatures on the two IDs match.

☐ Show the donor your identification, if requested.

☐ Enter the donor's ID number (i.e., social security number, badge number, or other employee number) on the chain of custody form. Record "donor did not provide ID number" on the chain of custody form if the donor does not provide an ID number. Enter the employer's name and address, the MRO's name and address, the drugs the specimen will be tested for, and the type of test on DOT chain of custody forms, except where this information is preprinted.

FIGURE 3–5. Collector's checklist.
Source: Adapted from U.S. Department of Transportation.[12]

- ☐ Ask the donor to remove unnecessary outer clothing (e.g., coat, jacket),* and to leave personal belongings (e.g., purse, briefcase) with the coat and jacket. Place all personal items in a secure place. Provide the donor with a receipt, if requested. Allow the donor to retain his/her wallet.
  [*Note: DOT states that, if the collection occurs during a DOT-mandated examination where the donor would otherwise disrobe, the specimen may be collected while the donor wears an examination gown.]

- ☐ Ask the donor to wash and dry his/her hands. If the donor refuses, record the refusal on the chain of custody form.

- ☐ From this point on, the donor is not allowed access to water fountains, faucets, soap dispensers, cleaning agents, or any other material that could be used to adulterate the specimen.

- ☐ Ask the donor to choose a wrapped/sealed collection container or specimen bottle.* Either the collector or the donor may unwrap or break the seal of the container or bottle.
  [*Note: The donor should choose from among several wrapped containers to help prevent accusations that the donor was given a contaminated container.]

- ☐ Advise the donor that a specimen of sufficient quantity (60 ml) is required.

- ☐ Ask the donor to provide the specimen in the privacy of the enclosure. If using a public rest room, remain in the rest room, but outside the enclosure, until the specimen is collected. (If an enclosed stall is unavailable, remain outside the closed rest room door.) Placard the rest room "not in use" during the collection process and keep others out. Ask the donor not to flush the toilet. (Fresh water flowing into the toilet could be used to dilute the specimen.)

- ☐ Ask the donor to keep the specimen in view at all times until it is packaged and sealed for shipment.

- ☐ Receive the specimen from the donor. Do not let the specimen out of your sight, or the donor's sight, until it is packaged and sealed for shipment.

FIGURE 3–5. Continued.

☐ Inspect the specimen for volume, color and temperature.* Measure temperature within 4 minutes.
[*Note: The specimen temperature may be read from either the collection container or the specimen bottle, so long as the reading is obtained within 4 minutes. If an operative thermometer is unavailable, the collector can estimate the temperature while holding the specimen bottle. If the specimen feels within normal range, the specimen should be processed. If not, the collection site supervisor should be contacted.[5]]

- If the volume is less than 60 ml, discard the specimen, ask the donor to drink fluids and collect another specimen.*
  [*Note: In non-DOT testing, separate specimens can be combined to reach 60 ml. The integrity of each partial specimen must be assured.]
- If the temperature is outside the acceptable range (90.5°-99.8°F, 32.5°-37.7°C), ask the donor to have his/her oral temperature taken. If the donor refuses or if his/her oral temperature is not within 1°C/1.8°F of the specimen, notify your supervisor or the employer representative to obtain authorization for a second, witnessed collection. Send both specimens to the laboratory.
- Note all visual or odor abnormalities of the specimen on the chain of custody form. If the specimen is obviously adulterated, request authorization from your supervisor or from the employer representative to conduct a second, witnessed collection. Send both specimens to the laboratory.

☐ If a separate collection container is used, pour the specimen into the specimen bottle while the donor watches. If a split specimen procedure is being done, pour the remainder into the second bottle.

☐ Place a cap on the specimen bottle and a tamper-proof seal over the cap and down the sides of the bottle in full view of the donor. The seal can be used as a label, or a separate label can be placed on the bottle. Write on the label the specimen number corresponding to the chain of custody form (unless it is preprinted on the label) and the collection date. Ask the donor to initial the seal or the label.

☐ Initiate the chain of custody block: Sign that you have received the specimen, and, on the next line, sign that you have released the specimen. If the specimen is shipped to the laboratory by a courier service, indicate this in the "Purpose of

FIGURE 3–5. Continued.

Change" column and identify the courier's name. If the specimen is put in a secure location at the collection site for pickup by the laboratory's courier, write "Secure Storage" in the "Purpose of Change" column; the laboratory will complete the "Received By" column upon receipt.

☐ Ask the donor to read and sign the donor's certification statement on the chain of custody form (and/or permanent record book). If the donor refuses to sign the certification statement, note this on the form (and/or in the record book). On the DOT chain of custody form, the donor should record his or her printed name, daytime phone number, and birth date.

☐ Sign the collector's certification statement on the form. On the DOT chain of custody form also record your printed name, the collection site location, and the collection date.*
[*Note: Corrections to the chain of custody form or specimen bottle label can be made, at the time of collection, by drawing a single line through the mistake, adding the correction, and initialing the change.]

☐ Distribute copies of the chain of custody form. Retain a copy at the collection site (copy 5 of the DOT form). Send a copy (copy 3 of the DOT form) to the MRO. With DOT forms, give copy 4 to the donor and send copy 6 to the employer.

☐ If there are problems during the collection, or if you have reason to believe the donor has adulterated the specimen, check with a supervisor or employer representative for further instructions. Note any unusual circumstances, behavior, or appearance on the chain of custody form or in the permanent record book.

AFTER COLLECTION

☐ Prepare the specimen for shipment.
- Make sure that copies of the chain of custody form have been distributed to the employer, donor, and/or MRO, as appropriate.
- Place the specimen and chain of custody form (copy 1 and 2 of the DOT form) in the shipping container, place a tamper-proof seal on the container, and initial and enter the collection date on the seal.
- Keep the specimens in a secure location until they are shipped.
- Temporary storage prior to shipping should be kept to a minimum.

FIGURE 3–5. Continued.

Split Samples

Certain drug testing programs include split specimen collections.* A split specimen is a single void into a collection container that is subsequently poured into two specimen bottles. One bottle is sent to the laboratory for analysis; the other is frozen and stored. The split specimen may be stored by the same laboratory, a different laboratory, the collection site, the employer, or the MRO. If the first specimen is laboratory-positive, the second specimen is then analyzed, usually at a different laboratory,† for the presence of the drug or metabolite found in the first specimen. The identity, integrity, and security of both specimens must be assured. The following procedures, in addition to those outlined in Figure 3-5, apply to the collection of split specimens, as described by DOT:

1. The donor chooses two sealed specimen bottles or a sealed collection container capable of holding more than 60 ml of urine.
2. If two specimen bottles are used, the donor fills one specimen bottle with 60 ml of urine and then fills the "split" specimen bottle with up to 60 ml of urine. There is no minimum volume requirement for the split specimen.
3. If a collection container is used, 60 ml is poured into one bottle and the remainder (up to 60 ml) is poured into the split specimen bottle. If the donor provides only 60 ml, the collection is valid and the collector should note in the remarks section of the chain of custody form that the donor was unable to provide a split.
4. Copy 7 of the DOT chain of custody form accompanies the split specimen to the laboratory. If a different chain of custody form is used, one with no copy for the split specimen, send a photocopy of the form with the split specimen.
5. If no urine is available for the split specimen, and the donor wants to wait until he/she can produce enough urine for both specimens, the 60 ml already collected should be processed and retained in a secure place for possible shipment. If the donor chooses not to wait, or if the donor cannot produce a specimen greater than 60 ml within 2 hours, the collector should note on the chain of custody form that a split specimen was not collected, and should send the originally collected specimen to the laboratory for analysis. If the donor does provide a

---

*The Omnibus Transportation Employee Testing Act requires split specimens on all collections conducted under its authority. Split specimens are otherwise an option that some employers use routinely or at the donor's request.

†NRC requires that a different laboratory analyze the split specimen.

specimen greater than 60 ml, the originally collected specimen and paperwork should be discarded, and a new chain of custody form should be initiated.[5]
6. When a split specimen is collected, the collector should note this on the chain of custody form along with the ID number corresponding to the split specimen.
7. The collector should try to use similar specimen identification numbers for each split specimen, for example, ID #9999A and ID #9999B.

The Omnibus Transportation Employee Testing Act[13] authorized DOT to develop procedures for split-specimen collections conducted under its authority. These procedures—which may differ from those in the 1989 DOT Procedures—were under development in 1992.

Parallel Collections

Some employers have two drug testing programs: one that meets federal requirements, and one that does not. The nonregulated program usually tests for additional drugs or at cutoff levels lower than the federal regulations. Separate specimens are submitted for analysis in parallel programs. The specimens must come from separate voids obtained during independent—for example, consecutive—collections. Nonregulated tests should not use chain of custody forms that reference the federal regulations and procedures.

Problem Solving

Donor Not Reporting to Collection Site
If the donor does not arrive at a scheduled collection, the collector should notify the employer and request further instruction.

Donor Not Bringing Photo Identification
If the donor brings no photo identification to the collection site, the collector should attempt to confirm the donor's identification through some other means, for example, by the employer or the employer's representative if he/she accompanied the donor to the collection site, or by contacting the employer's personnel office or the donor's supervisor. Without confirmation, the collector cannot proceed with the collection unless the donor is self-employed and has no supervisor. In this case, the collector should notify the collection site supervisor and record on the chain of custody form that positive identification is unavailable. The collector then obtains from

the donor two items of identification bearing his/her signature, and proceeds with the collection. When the donor signs the chain of custody form (or permanent record book), the donor's signature should be compared with the signatures on his/her items of identification. If the signatures do not match, the collector should note on the chain of custody form, "signature identification is unconfirmed."[5]

Monitored Collections
If a stall or other partially private enclosure is used for a nonwitnessed collection, DOT requires that a collector of the same sex as the donor, or a medically certified or licensed collector of either sex, monitor the collection from outside the stall.

Witnessed Collections
The donor urinates without being observed unless there is reason to believe that adulteration may occur. In the latter case, the collection should be directly observed—that is, the witness watches the urine pass from the donor's body to the container. The witness can be the collector or a different collection site staff person, but must be of the same sex as the donor. The chain of custody form should be annotated to indicate that the collection was witnessed, and should include the witness' name and signature.

DOT and NRC allow witnessed collections in only the circumstances listed in Figure 3-6. In the first two circumstances—abnormal temperature and clear evidence of adulteration—the collector must collect a second, witnessed specimen. The collector should annotate the chain of custody forms to reflect that both specimens were collected at the same appointment. Both specimens are sent to the laboratory for analysis. NRC allows testing to the limits of detection when adulteration is suspected. In circumstances 3 and 4 of Figure 3-6—the donor's last specimen was dilute, or the donor previously tested positive—the employer is responsible for notifying the collection site in advance if a witnessed collection is warranted. Federal procedures allow the collection site to check specimens for appearance and temperature, but (except for NRC regulations) do not allow the collection site to test the specimen beyond this, for example, for pH, specific gravity, or creatinine.

DOT requires that the collector obtain authorization from his/her supervisor or the employer representative before conducting a witnessed collection. Figure 3-7 is a form for recording the reason and authorization for conducting a witnessed collection. Figure 3-8 is a form that can be used to record the donor's consent to undergoing witnessed collection. Such forms

DOT and NRC regulations define each of the following as a "reason to believe that a particular individual may alter or substitute the specimen."

1. The urine specimen's temperature is outside the normal temperature range, and:

    a. The donor refuses to have his/her oral body temperature measured;
    or,
    b. The donor's oral body temperature varies by more than 1°C/1.8°F from the specimen's temperature.

2. Collection site personnel observe evidence clearly and unequivocally indicating an attempt to substitute or adulterate the specimen (substitute urine in plain view, blue dye in specimen, specimen smells of bleach, etc.).

3. The individual's last provided specimen was determined by the laboratory to have a specific gravity of less than 1.003 *and/or** creatinine concentration below .2g/L; or,

4. The specimen is being collected as a follow-up or return-to-work test under DOT/NRC rules for an individual who has previously had a verified positive DOT/NRC drug test.

NRC requires an immediate second collection under direct observation in each of the above circumstances. DOT requires an immediate second collection under direct observation in circumstances 1 and 2. DOT permits the employer to require that the next collection be conducted under direct observation in circumstances 3 and 4. (Circumstances 3 and 4 are not the basis for conducting the next DOT drug test. DOT instead permits a witnessed collection at the next test that would otherwise occur.)

*The DOT Procedures state "and." The NRC regulations state "or."

FIGURE 3-6. Circumstances for witnessed collections.

AUTHORIZATION FOR WITNESSED COLLECTION

_____   _____
Donor's Name   Date

_____   _____
Donor's ID Number   Collector's Name/Signature

REASON FOR WITNESSED COLLECTION

☐ Temperature outside range of 90.5°-99.8°F./32.5°-37.7°C.

    Urine Temperature: _____

                                               ☐ Individual declined to have
    Examinee's Temperature: _____   oral temperature taken.

☐ Collection site personnel observe evidence clearly indicating that the examinee attempted to substitute or adulterate the specimen.

(Describe evidence:) _____

_____

_____

☐ Last specimen had a specific gravity below 1.003 and a creatinine concentration below 0.2 g/L.

☐ The donor previously had a verified positive drug test.

APPROVAL FOR WITNESSED COLLECTION

Supervisor
OR
Employer Representative: _____ (print)

                                         _____ (signature)

                                         _____ (date)

FIGURE 3-7. Authorization form for a witnessed collection.

```
                    CONSENT FOR WITNESSED COLLECTION


    _____
    Donor's ID Number


    WITNESSED COLLECTION AUTHORIZED BY:

    _____ at _____ on _____.
         Name and Title              Company Name      Date/Time

    AUTHORIZATION TAKEN BY:

    _____ by _____.
    Collection Site Worker's Name/Signature   Telephone, Fax, Written Notification


    I, _____, have been notified and agree to my em-
         Donor's Name, Printed

    ployer's request that my urine specimen collection be witnessed by

    an employee of _____ on _____.
                    Collection Site's Name    Collection Date


                         _____ (Donor's Signature)
                         _____ (Date)
```

FIGURE 3–8. Consent form for a witnessed collection.

serve a risk-management function. Their use is not required by the federal drug testing regulations.

When a same-sex collector is needed but is not immediately available, the collector should notify the employer and the collection should be postponed until a same-sex collector is available.

Less Than 60 Milliliters of Urine
At least 60 ml of urine must be collected. (Revisions to federal regulations may lower the minimum volume to 30 ml.[4]) DOT requires a single void

of at least 60 ml. In DOT-mandated testing, if the donor cannot provide 60 ml from a single void, the collector should:

1. Discard the specimen and the collection container/specimen bottle used for the void;
2. Encourage the donor to drink reasonable quantities of fluids and remain at the collection site;
3. After a reasonable time ask the donor to attempt to provide another specimen.

In non-DOT tests, the collector may combine specimens from separate voids to reach 60 ml. In doing so, the collector should measure the temperature of each specimen when it is produced, should seal and unseal partial specimens while the donor watches, and should note on the chain of custody form that separate specimens were combined.

In DOT-mandated post-accident or reasonable cause testing, the donor should remain at the collection site for up to 8 hours until he/she provides a 60 ml specimen. If a 60 ml specimen is not provided within 8 hours, the collector should notify the employer that the collection was incomplete, and should document the time and circumstances.

In tests that are not post-accident or reasonable cause, the collection can be rescheduled or canceled at the employer's discretion if the donor cannot provide a 60 ml specimen within a reasonable time. A "reasonable time" is defined by the employer. DOT recommends 1–2 hours, and allows up to 8 hours.

If the donor cannot provide a 60 ml specimen within an 8-hour period or at a rescheduled DOT-mandated collection, the employer's MRO then refers the individual for a medical evaluation to determine if the inability has a physiologic basis or constitutes a refusal to provide a specimen. (DOT makes an exception for preemployment testing; the employer need not make such a referral if the employer does not wish to hire the individual.) Certain medical conditions may cause urine retention or difficulty in initiating micturition. Prescription and over-the-counter medications possessing anticholinergic properties—for example, atropine, phenothiazine, tricyclic antidepressants, and antihistamines—may also prolong the process. For most individuals, drinking up to 1 liter of water safely facilitates urination. Drinking larger quantities of water can cause water intoxication.[14] The symptoms of acute water intoxication include confusion, slurred speech, an unsteady gate, and, in severe cases, lethargy, confusion, and seizures.

Donor Requiring Medical Attention

If the donor needs medical attention—for example, in post-accident testing, the drug test collection must not delay that attention. If injury precludes collection of a fresh specimen—for example, the donor is in surgery, or has an indwelling catheter—the employer's records should reflect why testing was required, the steps taken to obtain a specimen, and the reason why collection was not possible.

Failure to Cooperate

It should be made clear to the donor that failure to cooperate with the procedures may have the same consequences as refusing to undergo the test. The collector should document noncooperation in the permanent record book or on the chain of custody form, and should notify the employer.

The Medical Examiner's Certificate and Drug Testing

The Federal Highway Administration (FHWA) regulations prohibit covered drivers from using certain drugs.[15] FHWA's drug testing regulations (Subpart H of 49 CFR § 391) authorize a periodic DOT drug test for covered drivers at the time of their biennial examination. (When conducting random testing at a 50 percent rate, the employer can discontinue periodic testing.) The medical examiner conducting the biennial examination determines the driver's physical qualification, and can do so with or without the results of a drug test. The medical examiner records the medical findings and determination on a form that has the statements shown in Figure 3-9. One of three scenarios may occur with respect to drug testing and the determination of physical qualification:

Controlled Substance Testing

☐ Controlled substances test performed

    ☐ In accordance with Subpart H.

    ☐ Not in accordance with Subpart H.

☐ Controlled substances test NOT performed.

FIGURE 3-9. Format for recording the drug test's status in a Federal Highway Administration medical examination.

- *In accordance with Subpart H.* If the medical examiner uses the DOT drug test result as part of the evaluation of the driver's physical qualification, the medical examiner checks the boxes: "Controlled substances test performed" and "In accordance with Subpart H."

- *Not in accordance with Subpart H.* If the medical examiner uses the result of a drug test performed with procedures other than those of Subpart H as part of the medical evaluation, the medical examiner checks the boxes: "Controlled substances test performed" and "Not in accordance with Subpart H."

- *Controlled substances test NOT performed.* If the medical examiner does not consider the result of a drug test as part of the medical examination, the medical examiner checks on the medical examination form: "Controlled substances test NOT performed."

The employer, not the medical examiner or the MRO, is responsible for ensuring that drivers are physically qualified and have been tested for controlled substances in accordance with the federal regulations.

AFTER COLLECTION

Packaging Standards

Containers for shipping urine specimens to the laboratory can be obtained from laboratories and from many of the courier services that transport specimens. Shipping containers are designed to prevent specimen damage, and should include the following components:

1. A watertight inner container;
2. A watertight outer container;
3. An absorbent material placed between the inside and outside containers. If multiple inner containers are placed in an outer container, they should be individually wrapped to prevent contact between them. The absorbent material must be sufficient to absorb the entire contents of all inner containers.

Some packaging systems include a second outer container, usually made of cardboard. Chain of custody documents should be placed between the inside and outside containers, so that that they will not be damaged by a potential spill.

Upon tape sealing the container, a collection site worker signs or initials his or her name and enters the date specimens were sealed in the shipping containers. The collection site worker also ensures that proper documentation is attached to each container sealed for shipment.

Specimen Security

Each specimen must be held in a secure place, preferably locked, before transport to the laboratory. Temporary storage should be kept to a minimum. If transport is delayed for several days, the specimen should be kept refrigerated (4°C) if possible. The storage place should be under constant supervision to prevent tampering. Any transfer of the specimen after packaging but before shipment should be noted on the collection site's copy of the chain of custody form. The number of people handling the specimen should be minimized.

Transport to the Laboratory

Specimens should be sent to the laboratory within 24 hours of collection. Specimens can be sent by U.S mail, but are more often sent by overnight courier or are picked up by the laboratory's courier. Third-party couriers offer quick service and maintain records that document deliveries. A specimen should never be turned over to the donor for storage or for transport to the laboratory.

Chain of Custody Errors

The most frequent procedural errors in collections involve the chain of custody forms. Certain errors may make the test invalid (unusable). For example, if the specimen I.D numbers on the chain of custody form and specimen bottle differ, the test is invalid. Certain other errors that can be corrected after the collection may also invalidate the test.[16] Correctable errors include:

1. Omitting the donor ID number from the form, unless "refusal of donor to provide number" is stated in the remarks section.
2. An incomplete chain of custody block (minimum: two signatures, shipping entry, dates).
3. Omitted collector's signature from the certification statement.

The collection site can correct any of these errors by completing a written statement regarding the error. Figure 3-10 illustrates a format for such statements. Actual statements should be typed on collection site letterhead,

I, _____, am an employee or authorized agent of
     Collector's Name

_____. On _____, I
  Collection Site Name and Address   Collection Date

omitted _____ from the drug test
            Explanation of Omission

custody and control form _____. I am submitting this affidavit
                  ID Number

to correct and amend the form.

               _____ (Collector's Name, Print)

               _____ (Collector's Signature)

               _____ (Date)

(This section is optional)

Sworn and subscribed before me

this _____ day of _____, 19____.

_____   (SEAL)

Notary public of: _____

My commission expires: _____

(Send the completed affidavit to the laboratory.)

FIGURE 3-10. Affidavit for omissions from the chain of custody form.

and should be notarized. They are usually submitted to whomever has requested the correction (usually the laboratory, less often the MRO).

Drug Testing Records

The collection site maintains either a copy of each chain of custody form or a bound book (see the "Supplies" section of this chapter) in which identifying data on each specimen are permanently recorded. Drug testing records should be stored separately from other employer-sponsored medical records, thereby helping to preserve the distinction between drug tests and medical examinations. DOT agency regulations require employers to maintain drug test records for up to 5 years for positive results and 1 year for negative results. Equivalent retention periods for collection site records are reasonable.

BREATH AND BLOOD ALCOHOL TESTING

Breath Testing

There are two types of breath-testing devices: (1) the preliminary breath-testing device, a hand-held device often used as a screening tool; and (2) the evidentiary breath testing device, generally a larger instrument that can provide results allowed as evidence in legal proceedings. In the breath alcohol test, the individual blows into an instrument that analyzes the expelled breath and provides a reading in terms of blood alcohol concentration (BAC). Evidentiary breath-testing devices provide a reasonably accurate measure of BAC, at least if they meet National Highway Traffic and Safety Administration specifications. These devices require well-trained collection site personnel and rigid adherence to a quality control program.

Alcohol breath tests must be delayed at least 15 minutes if any source of mouth alcohol (e.g., breath fresheners) is ingested or regurgitated (e.g., by vomiting or burping). The collector should help ensure that each specimen that undergoes preliminary breath testing comes from the end, rather than the beginning, of exhalation. Evidentiary breath-testing devices are designed to automatically measure end-expiratory air. The NRC regulations require two breath alcohol screening tests taken from 2–10 minutes apart. If the result of each test is within 10 percent of the average of the two measurements, the screening test is considered accurate. If the two tests do not agree, NRC requires repeating the breath tests on a different evidentiary-grade breath-testing device. NRC also allows the donor the option of submitting a blood specimen for alcohol analysis if breath testing indicates a BAC at or above 0.04 percent (i.e., 4 mg of alcohol per ml of blood or per 210 L of breath).

Breath tests, unlike blood and urine tests, provide an immediate response, so that follow-up action can be instituted promptly. They also do not require elaborate chain of custody procedures for collection or laboratory analysis, since no specimens are sent to a laboratory for analysis. On the other hand, breath testing requires specific training and quality control measures to ensure proper administration of the test and calibration of the device. Evidentiary breath-testing devices are expensive to purchase ($400–$5,000) and to maintain.

Another problem with breath testing is that no sample remains after the test for retesting in the event of a challenge. For this reason, some employers authorize a blood alcohol test when the breath test result exceeds the cutoff.

Blood Testing

Blood testing is the most reliable, and the most physically invasive, form of alcohol testing. Blood specimen collection is a medical procedure, and should be performed only by a qualified professional or technician and only at a suitable physical location. The skin is cleaned with a nonethanol cleanser, for example, Betadine®. (Ethanol on the skin can, in theory, contaminate the specimen.) Isopropyl alcohol cleaners should also be avoided because the donor may (wrongly) conclude that they significantly raise the blood alcohol measurement. The blood alcohol specimen is collected in a tube containing sodium fluoride and potassium oxalate. Chain of custody and packaging procedures are similar to those of urine testing. The donor remains with the specimen until it is completely labeled and sealed. A unique identification number is placed on the tube. An initialed and dated security seal is placed over the top and down the sides of the tube.

If both urine and blood are collected for nonregulated testing, the same chain of custody form can be used. DOT urine specimens and blood alcohol specimens need separate chain of custody forms.

SUMMARY

Proper specimen collection is crucial to the success of any drug testing program. The specimen's integrity must be assured, or the test itself cannot be valid. The donor's privacy must be respected, or the entire testing program will be subject to criticism. Adherence to well-designed collection procedures is therefore essential.

References
1. Department of Health and Human Services (DHHS). Mandatory guidelines for federal workplace drug testing programs. *Fed. Reg.* 1988;53(April 11):11,970–11,989.

2. Nuclear Regulatory Commission. Fitness-for-duty Programs. *Fed. Reg.* 1989;54(June 7):24,468–24,508.
3. U.S. Department of Transportation. Procedures for transportation workplace drug testing programs. *Fed. Reg.* 1989;54(Dec. 1):49,854–49,884.
4. Specimen collection and reporting results. In B. S. Finkle, R. V. Blanke, and J. M. Walsh (eds.), *Technical, Scientific, and Procedural Issues of Employee Drug Testing.* Rockville, Md.: National Institute on Drug Abuse; 1990:21-24.
5. U.S. Department of Transportation. *Specimen Collection Workbook.* Washington, D.C. September 1991.
6. Centers for Disease Control. Update: Universal precautions for prevention of transmission of human immunodeficiency virus, hepatitis B virus, and other bloodborne pathogens in health-care settings. *Morbidity and Mortality Weekly Report* 1988;37:377–387.
7. Autry, J. H. *Advisory 4: Permanent Record Book and MRO Review.* Rockville, Md.: National Institute on Drug Abuse; October 25, 1991.
8. 49 CFR § 40.25(f)(22)(ii).
9. 29 CFR § 1910.20(c)(10)(i).
10. Department of Health and Human Services (DHHS). *Urinalysis Collection Handbook for Federal Drug Testing Programs.* Washington, D.C. 1988.
11. Federal Aviation Administration. *Standard Specimen Collection Procedures for Department of Transportation Anti-Drug Programs.* Washington, D.C. 1990.
12. U.S. Department of Transportation. *Urine Specimen Collection Procedures Guide.* Washington, D.C. 1990.
13. Omnibus Transportation Employee Testing Act of 1991. Pub law 102-143, 102nd Congress, October 28, 1991.
14. Klonoff, D. C., and A. H. Jurow. Acute water intoxication as a complication of urine drug testing in the workplace. *JAMA* 1991;265:84–85.
15. 49 CFR § 391.41(b)(12).
16. U.S. Department of Transportation. *Operating Guidance for DOT Mandated Drug Testing Programs.* Washington, D.C. May 29, 1991.

4

Forensic Laboratory Drug Testing

Alan B. Jones, Ph.D., and
Michael A. Peat, Ph.D.

LABORATORY SELECTION

Selection of a quality laboratory is essential. Workplace drug testing demands the utmost confidence in the laboratory's reliability. Laboratory costs are often the single greatest expense in conducting testing. Thus, the laboratory should be carefully selected based on the considerations presented in Figure 4-1.[1] Among the most significant of these are certification, for example, certification by the Department of Health and Human Services (DHHS), or by the College of American Pathologists (CAP).

Specimens from federal employees and from employees in certain federally regulated industries must be analyzed at DHHS-certified laboratories according to the procedures in the DHHS Mandatory Guidelines. DHHS certification is achieved through on-site inspections and periodic performance testing, as described in the DHHS "Mandatory Guidelines for Federal Workplace Drug Testing Programs."[2] A list of DHHS-certified laboratories is published in the *Federal Register* on or about the first Friday of every month.* This list is updated at each publication to indicate additions, deletions, suspensions, and withdrawals from the program. As of April 1992, there were 85 DHHS-certified—also referred to as NIDA-certified—laboratories. National Institute on Drug Abuse (NIDA) certification of one

---

*This list can be obtained by contacting the Division of Applied Research, National Institute on Drug Abuse, Parklawn Building, 5600 Fishers Lane, Rockville, Md, 20857.

> 1. Is the laboratory licensed and/or accredited to perform drug analyses?
> 2. What drugs are included in the screening test(s) and the confirmation test(s)?
> 3. What is the cost of testing?
> 4. What is the turnaround time?
> 5. Is a chain of custody procedure available?
> 6. What supplies (urine containers, etc.) are provided?
> 7. What is the laboratory's report mechanism and how is confidentiality ensured?
> 8. How long can positive specimens be stored at the laboratory?
> 9. Can the laboratory offer support (technical testimony) at hearings or in court-related cases?

FIGURE 4-1. Laboratory selection questions. (When testing is conducted in accordance with the DHHS Mandatory Guidelines, the answers to many of these questions are defined by the Guidelines.)

laboratory within a multilaboratory company does not imply certification of the company's other laboratories. Furthermore, even within a NIDA-certified laboratory, specimens are not necessarily analyzed according to DHHS procedures unless the client explicitly requests this.

The CAP certification program is organized under the joint sponsorship of the CAP and the American Association of Clinical Chemistry (AACC).[3] This certification process is similar to NIDA certification in that to qualify for inspection and accreditation by CAP, a laboratory must successfully complete three consecutive sets of proficiency samples. Accredited laboratories must maintain proficiency on an annual basis to retain certification. One of the major differences between the NIDA and CAP certification programs is that NIDA requires the use of a two-tiered testing methodology—an immunoassay screening test with gas chromatography-mass spectrometry (GC/MS) confirmation testing—whereas CAP does not dictate which analytic methodology must be used.

Each laboratory should offer technical support to the client. The nature of the technical support varies from laboratory to laboratory, and should include a consultant service to assist in interpretation of specific laboratory results. Most laboratories have on their staff a toxicologist trained in forensic urine drug testing who can help in such situations. This technical support should extend to presenting, explaining, and defending results in legal proceedings that might arise from a drug test result.

Other important considerations in laboratory selection include efficient means of shipping specimens to the laboratory, clearly written instructional materials, and, if desired, the availability of collection services through the laboratory.

ANALYTICAL PROCEDURES

The forensic urine drug testing laboratory uses analytical methodologies that have been developed specifically for this arena. A laboratory flow chart depicting the sequence of these procedures is shown in Figure 4–2.

Upon receipt of a specimen by the laboratory, the laboratory personnel first check the packaging and the accompanying paperwork (the external chain of custody form) for completeness, legibility, agreement of specimen identification numbers on bottle and paperwork, and so on. An aliquot of the specimen is then removed for initial testing. The receipt, identification, and removal of an aliquot is know as *accessioning*. Accompanying this aliquot is another document generated by the laboratory that tracks the move-

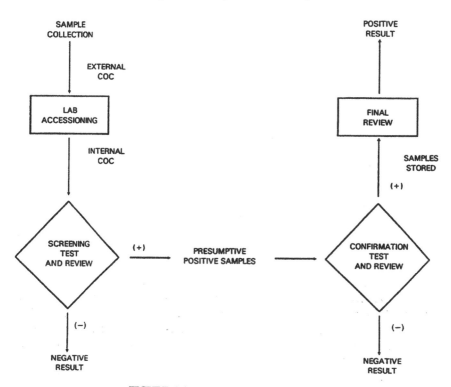

FIGURE 4–2. Laboratory flow chart.

ment of the aliquot through its processing. This document is commonly referred to as an *internal chain of custody document*. Its format varies from laboratory to laboratory; however, its purpose is the same everywhere: to produce a paper trail to record everything and everyone who has contact with the aliquot during analysis.

Each specimen undergoes an initial test that serves to eliminate specimens that are likely negative from further consideration. The initial tests that are conducted on the aliquot are generally immunoassay-based. The NIDA-certified laboratories are required to use an FDA-approved (Food and Drug Administration) immunoassay test as the initial test. Several immunoassays are available from commercial manufacturers: the enzyme multiplied immunoassay marketed under the trade name EMIT by Syva Corporation; the fluorescence polarization immunoassay marketed under the trade name TDx by Abbott Laboratories; the radioimmunoassays marketed under the trade names Abuscreen by Roche Diagnostics, Inc.; and Coat-A-Count by Diagnostic Products Corporation.

The basis of immunoassay testing is that each assay has an antibody that binds specifically to the drug or analyte for which the test is designed. The differences between the various commercial immunoassay methodologies lies in the manner in which the antibody has been produced, the manner in which the antibody has been incorporated into the test system—that is, suspended in media, bound to a solid surface, and so on—and the manner in which binding of the antibody to the drug is detected. The tests are configured so that the magnitude of the response (i.e., the amount of binding between analyte and antibody) is proportional to the analyte concentration. Thus, these assays are semiquantitative in that they are constructed to differentiate between ranges of analyte concentrations.

Immunoassays are relatively specific in that the antibodies do not bind to compounds unless they are very similar to the analyte for which the test is designed. There is an immunoassay specific for each drug or class of drugs. Associated with each immunoassay is a cutoff value that is used to determine if a given specimen is to be classified as negative or presumptive positive. If the specimen's concentration is found to be at or above the preestablished cutoff, the specimen is considered a presumptive (or provisional) positive and the specimen is further tested by confirmatory procedures. Conversely, if the concentration of a test specimen is found to be below the cutoff, the result will be reported as negative. Thus, in forensic urine drug testing, negative does not mean "no drug present" or "undetected" but instead means the apparent concentration of analyte was less than the preestablished cutoff concentration for that assay. As noted above, each immunoassay has a cutoff unique to that assay. Table 4-1 lists the immunoassay cutoff concentrations established for federally mandated testing.

Most laboratories offer a variety of drug testing panels, some of which use cutoff levels that differ from those of the NIDA Guidelines. These panels do not meet the requirements of the NIDA Guidelines, even though they may be conducted at a NIDA-certified laboratory. Furthermore, the laboratory must follow the standard operating procedures approved during NIDA certification if the test is to meet the requirements of the NIDA Guidelines. The marijuana metabolites test is the most frequent test offered at non-NIDA cutoff levels. Laboratories typically offer lower immunoassay cutoff levels for this test, for example, 20 or 50 ng/ml.

In testing conducted according to the NIDA Guidelines, no further tests are conducted on any specimen identified as negative in the screening test. The results are reviewed by a certifying scientist at the laboratory and, if everything is in order on that review, the result is reported as negative. If the specimen's screening test result is positive, then a new aliquot of the specimen is obtained and a new internal chain of custody is initiated for this aliquot. The aliquot is transferred to the area of the laboratory responsible for confirmation testing. Confirmation testing is conducted *only* for those drugs and/or metabolites that were shown to be present at or above the cutoff level in the immunoassay screening test.

The NIDA Guidelines define the confirmatory test as "a second analytical procedure to identify the presence of a specific drug or metabolite which

TABLE 4-1 NIDA Drug Testing Cutoff Concentrations

Drugs	Initial (Screening) Cutoff ng/ml	Confirmatory (GC/MS) Cutoff ng/ml
Marijuana metabolites	100[a]	15[b]
Cocaine metabolites	300	150[c]
Opiate metabolites	300[d]	
Morphine		300
Codeine		300
Phencyclidine	25	25
Amphetamines	1000	
Amphetamine		500
Methamphetamine		500

[a] Proposed to change to 50 ng/ml.
[b] Assayed as 11-nor-delta-9-THC-9-carboxylic acid (a THC metabolite).
[c] Assayed as benzoylecgonine (a cocaine metabolite).
[d] 25 ng/ml if the immunoassay is specific for free morphine.

is independent of the initial test and which uses a different technique or chemical principle from that used in the initial test." The NIDA Guidelines specify that this second technique is GC/MS. The GC/MS procedures that are currently used employ an extraction step in which the drug or analyte in question is separated from the urine via a procedure designed specifically for that analyte. Most GC/MS procedures then chemically change the analyte to form a derivative. This derivative is injected into the gas chromatograph (GC) and examined by the mass spectrometer (MS), which serves as a detector for the GC. The magnitude of the response from the GC/MS is directly proportional to the amount of material injected and is thus proportional to the concentration of analyte in the original specimen.

GC/MS is highly sensitive and can detect most drugs in low nanograms per milliliter concentrations. GC/MS is also very specific; it examines the specimen for a single chemical entity, either the drug or the drug metabolite. The major drawbacks to GC/MS are the time it takes to prepare the sample before testing and the need for more highly trained staff to perform the analyses.

It is only after a specimen has tested positive in both the initial and confirmatory tests that it is reported as laboratory positive. NIDA-certified laboratories report the result as positive or negative and, if positive, for which drug. The quantitative level(s) in a positive specimen can be obtained upon request. Also, per the NIDA Guidelines, those specimens identified as positive by the laboratory are stored at $-20°C$ and retained for at least one year; if legal proceedings are in progress, storage continues as long as required by those proceedings.

In federally mandated testing, the laboratory will honor a request for reanalysis only if it is presented from the Medical Review Officer (MRO). Upon receipt of such a request, the laboratory will retrieve the specimen from storage, remove another aliquot, initiate an internal chain of custody document for this aliquot, and transfer the aliquot to the confirmation testing area. In reanalysis, only the GC/MS testing is repeated (immunoassay testing is skipped) and the criteria for calling the specimen positive or negative change. In reanalysis, the laboratory need only show the presence of the analyte in the tested aliquot to call the test positive. The cutoff concentrations do not apply for several reasons:

1. Drugs and metabolites degrade over time, even when frozen.
2. Drugs and metabolites precipitate when frozen and may not readily return into solution when thawed.
3. Drugs and metabolites adhere to the walls of the specimen containers.
4. Because of analytical variability, results can fall above or below the cutoff when the true concentration is near the cutoff.

These same criteria are applied to analyses of split specimens. To call a split specimen positive, the laboratory need only show the presence of the particular analyte for which the specimen is being tested.

THE DRUGS OF ABUSE

There are numerous drugs presently abused. The following sections address the drugs for which testing is most commonly conducted.

Marijuana

Marijuana is the processed leaf material from the plant *Cannabis sativa*. It contains a great variety of chemical entities, among which is a class of compounds unique to the plant known as cannabinoids.[4] To date there have been 62 different cannabinoids isolated and identified from the plant material. Delta-9-tetrahydrocannabinol (delta-9-THC) is the primary compound responsible for the psychotropic activity associated with the use of marijuana.[5] Marijuana is typically smoked in cigarettes, where the delta-9-THC is volatilized and trapped in the smoke particles, and thus delivered deep into the lungs where rapid absorption into systemic circulation occurs. Once absorbed, delta-9-THC is rapidly distributed to other tissues of the body. The compound is extensively metabolized, and its metabolites are primarily eliminated via the feces and urine, with little or no parent compound being eliminated via either route.[5]

The metabolic profile varies from individual to individual, especially as the level of use varies. The delta-9-THC metabolite appearing in the greatest concentration is usually 11-nor-delta-9-THC-9-carboxylic acid, which in itself has no pharmacological activity but is the target compound toward which both the screening and confirmation tests are directed. Because the immunoassay antibody binds to some of the other metabolites,[6] the NIDA cutoff concentrations for screening (100 ng/ml) and confirmation (15 ng/ml) differ significantly. Furthermore, because the ability of the antibodies to bind the various metabolites varies between manufacturers, a marijuana-positive urine specimen may yield different results when tested by different immunoassays. Because the GC/MS procedure is specific for the 11-nor-delta-9-THC-9-carboxylic acid metabolite, there should be only normal laboratory variation in the results of repeated analyses, even when performed at different laboratories.

Cocaine

Cocaine is an alkaloid from the coca plant, *Erythroxylon coca*.[7] It is usually obtained as cocaine HCl (hydrochloride), but those who smoke the drug remove the HCl to prepare the "freebase" or "crack" form that is smoked. Various purities of this drug are found, ranging from coca paste—an intermediate product during refinement, containing 50–95 percent cocaine—to pure cocaine. The drug is usually smoked or snorted, but may also be ingested or injected.[8] When smoked or snorted, the cocaine is rapidly absorbed into systemic circulation via the pulmonary system or the nasal mucosa. Cocaine in the body is rapidly and extensively metabolized by liver and plasma enzymes, primarily to benzoylecgonine and ecgonine methyl ester. Both of these metabolites are excreted in the urine along with some unmetabolized cocaine. The half-lives for these three compounds—benzoylecgonine, ecgonine methyl ester, and cocaine—are 7.5 hours, 3.6 hours, and 1.5 hours, respectively.[9]

The immunoassay and GC/MS procedures are both directed toward benzoylecgonine, the compound with the longer half-life. Benzoylecgonine can usually be detected for 2–3 days after a single dose. The immunoassay screening procedures respond to several cocaine metabolites; thus, the screening cutoff concentration is typically 300 ng/ml. The GC/MS confirmation procedures are specific for benzoylecgonine and typically have cutoff concentrations of 150 ng/ml. Several studies present data that attempt to correlate cocaine dosages with urine concentrations over time.[9,10]

Amphetamine and Methamphetamine

Amphetamine and methamphetamine are among the central nervous system stimulants and hallucinogens that are collectively referred to as *amphetamines*. These drugs are used rarely to treat narcolepsy, attention-deficit disorder, depression, and obesity.[11] Because of the abuse risk, medical boards or organizations in several jurisdictions have formally determined that it is inappropriate to treat obesity with these drugs for more than a few weeks. Amphetamine and methamphetamine stimulate attention and prolong endurance; these characteristics make them particularly subject to abuse by truck drivers, athletes, and students.

Methamphetamine is more often abused in the United States, whereas amphetamine is more often abused in other countries. Methamphetamine use has been especially prevalent in southern California. Methamphetamine is often synthesized in clandestine laboratories, which produce a mixture of both the *d* (dextro) and *l* (levo) optical isomers. The *l* isomer, *l*-methamphetamine, is also contained in the over-the-counter decongestant, Vicks Inhaler®. The *l* isomer has primarily peripheral action and, unlike the *d* form, does not have high abuse potential. Because of these distinctions, it is important that the forensic drug testing laboratory be able to differentiate between the *d* and *l* isomers of methamphetamine.[12]

Certain drugs metabolize to methamphetamine and amphetamine.[11] For example, benzphetamine and selegiline metabolize to both methamphetamine and amphetamine; selegiline selectively metabolizes to the *l* isomers. Methamphetamine metabolizes to amphetamine, so that both may be found in the urine after methamphetamine use.

The excretion of amphetamine and methamphetamine is pH-dependent. A single therapeutic dose produces positive urine tests for about 24 hours. High-dose users may test positive for several days after use.[13]

Several immunoassays are available for screening of these compounds. To some degree, they all suffer from the problem of cross-reactivity. The EMIT kits, marketed by Syva, contain either a polyclonal or monoclonal antibody. The polyclonal kit cross-reacts strongly with other sympathomimetic amines such as ephedrine, pseudoephedrine, and phenylpropanolamine. The monoclonal kit, by comparison, is less cross-reactive to *l*-methamphetamine and to other sympathomimetic amines. The FPIA Amphetamine II kit marketed by Abbott Diagnostics has less cross-reactivity to the sympathomimetics than either the polyclonal or monoclonal EMIT kits. All of the immunoassay kits cross-react with phentermine, MMDA (5-methoxy-3,4-methylenedioxyamphetamine), and MDA (3,4-methylenedioxyamphetamine). It is important to note that none of the immunoassay kits are entirely specific for amphetamine and/or methamphetamine.

GC/MS procedures used for routine confirmation of the amphetamines are designed to identify amphetamine and methamphetamine. In late 1990, it was found that at very high temperatures during processing and at very high concentrations, ephedrine (in the hundred-thousands of ng/ml) can cause false-positive GC/MS tests for methamphetamine. It is unclear how this occurs—some have hypothesized that the ephedrine mimics the pattern produced by methamphetamine, or that the ephedrine is converted to methamphetamine. The problem can be resolved by careful selection of analytical conditions and by making it a prerequisite that a methamphetamine-positive specimen must also contain 200 ng/ml or greater of amphetamine in order to be reported as a positive result. (The latter criterion is currently required in federally regulated drug testing.[14]) In all of the false-positive reports for methamphetamine, amphetamine was either absent or, if present, was at concentrations much lower than 200 ng/ml.

Morphine and Codeine

Morphine and codeine are classified as opiate narcotics and are present in the opium poppy. Morphine can be converted (e.g., in clandestine laboratories) to heroin (diacetylmorphine). Heroin historically has been injected, but more recently, because of the dangers of sharing needles, there have been numerous reports of heroin use by smoking and snorting. When used by these routes it is often mixed with cocaine.

Morphine and codeine are both used as analgesics. Morphine is more often used in hospitals, as opposed to outpatient settings. Codeine is included in numerous analgesic medications, such as Tylenol #3®, and as an antitussive in cough tinctures. Generally these preparations are available by prescription only, although in some states an individual need only sign a pharmacist's log for certain codeine-containing medications that are controlled under Drug Enforcement Agency regulations.

The body metabolizes codeine to morphine, as shown in Figure 4–3.[15] Morphine is not, however, metabolized to codeine.[16] Both codeine and morphine may occur in urine following use of only codeine. Morphine in the urine may represent use of morphine or codeine, or ingestion of poppy seeds.

Ingestion of poppy seeds can potentially result in urine morphine concentrations exceeding 1000 ng/dl and codeine levels exceeding 300 ng/dl for short periods.[17,18] Distinguishing poppy seed ingestion from drug use is difficult and sometimes impossible. 6-Monoacetylmorphine (6-MAM) is a specific metabolite of heroin, and when found in the urine it is conclusive evidence of heroin use. However, 6-MAM is present in the urine for only a very short time (up to 12 hours) and at very low concentrations after a single dose of heroin. Thus, the absence of 6-MAM does not rule out heroin use.[19]

Several immunoassays are available for detecting opiates. Each is directed toward morphine, but each has different cross-reactivities to conjugated morphine, codeine, and the synthetic opiates, such as hydrocodone, hydromorphine, and oxycodone. The NIDA Guidelines require a 300-ng/ml cutoff for opiates, or, if the immunoassay is specific for free morphine, a 25-ng/ml cutoff. The NIDA Guidelines require a 300-ng/ml cutoff for each opiate at GC/MS testing. Generally these assays involve an acid or enzymatic hydrolysis of the urine and therefore detect "total" morphine and codeine. These assays at the aforementioned cutoffs detect opiates for several days after use.

Phencyclidine (PCP)

PCP is a hallucinogenic drug that has no legitimate use. It was first synthesized in 1956. Until recently, PCP was available as a veterinary anesthetic. It is metabolized to a number of metabolites, the primary one being 4-

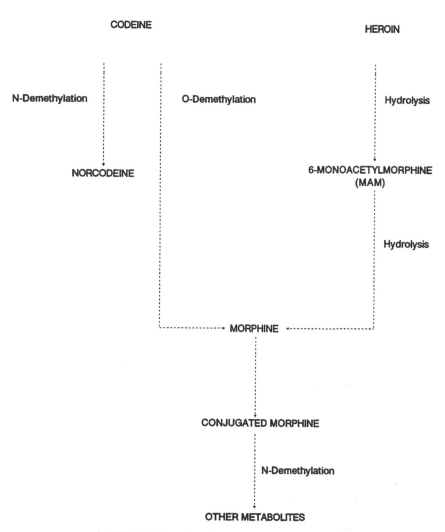

FIGURE 4-3. Codeine metabolized to morphine.

hydroxyphencyclidine. Like THC, PCP is lipid-soluble and there have been anecdotal reports that it is excreted very slowly over a number of days. The drug is usually smoked. It can be detected using a cutoff of 25-ng/ml for a number of days after use.[20] The commercially available immunoassays are designed to detect PCP but may also detect high concentrations of doxylamine or dextromethorphan. The GC/MS analysis is specific for PCP.

Barbiturates

Barbiturates are derivatives of barbituric acid and are central nervous system depressants. The most commonly abused barbiturates are short- and intermediate-acting agents such as pentobarbital, secobarbital, and amobarbital. Long-acting agents such as phenobarbital are rarely subject to abuse.

Barbiturate-containing medications are used for anesthesia, and for the treatment of insomnia and some convulsive disorders. Several barbiturate-containing medications are available over the counter; these include Donnagel® and Primatene P®. Barbiturates have been shown to be addictive and to have significant abuse potential. They are misused to counteract the stimulant effects of amphetamines and to potentiate the effects of heroin.

Immunoassays for barbiturates are generally designed to detect secobarbital, but will cross-react with amobarbital, pentobarbital, butalbital, and other barbiturates if they are present. GC/MS analyses are usually targeted at a panel of barbiturates consisting of amobarbital, butalbital, pentobarbital, phenobarbital, and secobarbital. Some laboratories also look for butabarbital. There are no uniform (e.g., NIDA) standards for barbiturate analyses. The immunoassay cutoffs are usually 300 ng/ml, and the GC/MS cutoffs are usually 200 ng/ml.

Benzodiazepines

The benzodiazepines are considered by many as the most prescribed drugs in the United States. The best known benzodiazepine drug is diazepam (Valium®); others include alprazolam (Xamax®), lorazepam (Ativan®), oxazepam (Serax®), and flurazepam (Dalmane®). They are primarily used as antianxiety and sedative-hypnotic drugs, and also as anticonvulsants and muscle relaxants.[21] The benzodiazepines have varying pharmacokinetic profiles, depending on their variations in chemical structure. Most are completely absorbed following oral administration and reach systemic circulation in the form of their active metabolite with peak plasma concentrations occurring from 0.5–8 hours postingestion. Their biotransformation and elimination half-lives vary from 2–3 hours to more than 50 hours. Some of the benzodiazepines have a common intermediate metabolite, nordiazepam, which undergoes further biotransformation. GC/MS analyses are usually targeted at nordiazepam and oxazepam. Like barbiturates, there are no uniform standards for benzodiazepine analyses. The immunoassay cutoffs are usually 300 ng/ml, and the GC/MS cutoffs are usually 50–100 ng/ml.

ALCOHOL

Alcohol analyses are perhaps the most commonly performed drug analyses. Accurate and precise methods are available for measuring alcohol in blood and breath samples. Breath alcohol is commonly converted to blood alcohol concentration using a ratio of 2,100:1. Urine and saliva specimens can also be tested for alcohol,[22] but these specimens, particularly urine, correlate less well with blood alcohol concentration (BAC).

BACs are usually expressed as percent weight/volume, that is, the grams of ethanol per 100 milliliter of blood, multiplied by 100 percent. BAC can be related to impairment.[23] In some states, a BAC of either 0.08% w/v or 0.100% w/v is considered proof of impairment for driving a motor vehicle.

Most forensic alcohol testing is done on expired breath, rather than blood, for many reasons. It is noninvasive, is rapid, and can be used in the field by nonmedical people. Although various issues have been raised to challenge the validity of breath alcohol measurement, there is no question that when the test is performed correctly it results in an accurate reading of the alcohol concentration.

The measurement of alcohol in blood is straightforward. In the forensic setting, specimen collection and laboratory quality control procedures are similar to those used in urine drug testing. The analysis may be performed by chemical, biochemical, or gas chromatographic assays.

Alcohol is metabolized at an average rate of 0.018 percent per hour; rates as low as 0.009 percent and as high as 0.03 percent have been reported. It is possible to back-extrapolate a BAC on an assumed metabolic rate, but care should be taken when doing so.[24] Factors such as delayed absorption may prevent an accurate estimation of BAC.

HAIR ANALYSIS

Various reports have looked at the detection of drug abuse via hair analysis.[25,26,27,28] There are, however, no peer-reviewed articles that validate hair/drug analyses. Furthermore, at a NIDA-sponsored conference in 1989, experts in the field of drug abuse concluded, "there is insufficient data available to recommend [hair analysis] in mass screening at this time." Federal drug testing regulations require that analyses be performed with urine samples, tested by an immunoassay test followed by GC/MS confirmation of positives. Hair, once dissolved in solution, may not produce sufficient concentrations for GC/MS testing.

Hair analysis also raises quality assurance problems. The required extractions are laborious and tedious, and do not appear to lend themselves reliably to the economics or technical skills of large-scale commercial laboratory work. Furthermore, the stability of drugs in hair is presently unknown. Contamination by recent passive exposure (e.g., sidestream smoke) is more likely than not. And, whether exposure to external agents, such as engine exhaust, may cause degradation of drugs in hair is unknown.[29] Finally, the equity of drug surveillance of hair must be addressed. Almost all people make urine; all have blood. But the amount of available hair differs markedly from person to person.

SUMMARY

The federal drug testing regulations specify that specimens be positive by both screening (immunoassay) and confirmation (GC/MS) assays before they are reported as laboratory-positive. The immunoassays are a mechanism to eliminate negative specimens from further consideration. The GC/MS procedures yield specific information on the kind and amount of drug present in specimens with positive immunoassay test results. A positive laboratory result indicates previous exposure to the drug. The pharmacology of the drug should be considered when interpreting potential explanations for its presence.

References
1. Warner, A., F. M. Hassan, and W. K. Fant. Drug abuse testing: Pharmacokinetics and technical aspects. *Am. Assoc. Clin. Chem. Therapeutic Drug Monitor.-Toxicol.* 1986;8:1–11.
2. Department of Health and Human Services (DHHS). Mandatory guidelines for federal workplace drug testing programs. *Fed. Reg.* 1988;53(April 11):11,970–11,989.
3. College of American Pathologists. *Standards for Accreditation. Forensic Urine Drug Testing Laboratories.* Northfield, Ill. 1990.
4. Turner, C. E., M. A. El Sohly, and E. G. Boeren. Constituents of *cannabis sativa L.* XVII. A review of the natural constituents. *J. Nat. Prod.* 1980;43:169.
5. Chiang, C. N., and G. Barnett. Marijuana pharmacokinetics and pharmacodynamics. In K. K. Redda, C. A. Walker, and G. Barnett (eds.), *Cocaine, Marijuana, Designer Drugs: Chemistry, Pharmacology, and Behavior.* Boca Raton, Fla.: CRC Press; 1989:113–124.
6. Jones, A. B., H. N. El Sohly, and M. A. El Sohly. Cross-reactivity of selected compounds in urine immunoassays for the major metabolite of delta-9-tetrahydrocannabinol. In: D. J. Harvey (ed.), *Marijuana—'84: Proceedings of the Oxford Symposium on Cannabis.* Oxford, United Kingdom:IRL Press; 1985:137.
7. Aldrich, M. R., and R. S. Barker. Historical aspects of cocaine use and abuse. In S. J. Mule (ed.), *Cocaine: Chemical, Biological, Clinical, Social and Treatment Aspects.* Cleveland: CRC Press; 1976:1–11.

8. Brown, R. Pharmacology of cocaine abuse. In K. K. Redda, C. A. Walker, and G. Barnett, (eds.), *Cocaine, Marijuana, Designer Drugs: Chemistry, Pharmacology, and Behavior.* Boca Raton, Fla.: CRC Press; 1989:39–51.
9. Ambre, J. The urinary excretion of cocaine and metabolites in humans: A kinetic analysis of published data. *J. Anal. Toxicol.* 1985;9:241–245.
10. Hamilton, H. E., J. E. Wallace, E. L. Shimek, P. Land, S. C. Harris, and J. G. Christenson. Cocaine and benzoylecgonine excretion in humans. *J. Forensic Sci.* 1977;22:697–707.
11. Goodman, A. G., L. S. Goodman, and A. Gilman. *Goodman and Gilman's The Pharmacological Basis of Therapeutics, 6th Edition.* New York: Macmillian; 1980:159–163.
12. Fitzgerald, R. L., J. M. Ramos, S. C. Bogema, and A. Poklis. Resolution of methamphetamine stereoisomers in urine drug testing: Urinary excretion of R(-)-methamphetamine following use of nasal inhalers. *J. Anal. Toxicol.* 1988;12:255–259.
13. Beckett, A. H., and M. Rowland. Urinary excretion kinetics of amphetamine in man. *J. Pharmacy Pharmacol.* 1965;17:628–639.
14. Autry, J. H. *Notice to All DHHS/NIDA Certified Laboratories.* Rockville, Md.: National Institute on Drug Abuse; December 19, 1990.
15. Stead, A. H. The use of pharmacokinetics and drug-metabolism data in forensic toxicology. In J. W. Bridges, and L. F. Chasseaud, (eds.), *Progress in Drug Metabolism.* London: Taylor and Francis, 1986;175.
16. Mitchell, J. M., B. D. Paul, P. Welch, and E. J. Cone. Forensic drug testing for opiates. II. Metabolism and excretion rate of morphine in humans after morphine administration. *J. Anal. Toxicol.* 1991;15:49–53.
17. El Sohly, M. A., and A. B. Jones. Morphine and codeine in biological fluids: Approaches to source differentiation. *Forensic Sci. Rev.* 1989;1:14–21.
18. Selavka, C. M. Poppy seed ingestion as a contributing factor to opiate-positive urinalysis results: the Pacific perspective. *J. Forensic Sci.* 1991;36:685–696.
19. Cone, E. J., P. Welch, J. M. Mitchell, and B. D. Paul. Forensic drug testing for opiates, I. Detection of 6-acetylmorphine in urine as an indicator of recent heroin exposure: Drug and assay consideration and detection times. *J. Anal. Toxicol.* 1991;15:1–7.
20. Hawks, R. L., and C. N. Chiang. Examples of specific drug assays. In R. L. Hawks, and C. N. Chiang, (eds.), *Urine Testing for Drugs of Abuse.* Rockville, Md.: National Institute on Drug Abuse, 1986;84–112.
21. Miller, N. S., and M. S. Gold. Benzodiazepines. In: A. J. Giannini and A. E. Slaby, (eds.), *Drugs of Abuse.* Oradell, N.J.: Medical Economics Books, 1989;59–82.
22. Kadehjian, L. Urine alcohol testing: Valid and valuable. *Syva Monitor* 1991;9:9–13.
23. Dubowski, K. M. Alcohol determination in the clinical laboratory. *Am. J. Clin. Pathol.* 1980;74:747–750.
24. Wu, A. H. B. Pharmacokinetic aspects of ethanol for medicolegal purposes. *Am. Assoc. Clin. Chem. Therapeutic Drug Monitor.-Toxicol.* 1991;12:7–13.
25. Graham, K., G. Koren, J. Klein, J. Schneiderman, and M. Greenwald. Deter-

mination of gestational cocaine exposure by hair analysis. *JAMA* 1989;262:3,328-3,330.
26. Marigo, M., F. Tagliaro, C. Poiesi, S. Lafisca, and C. Neri. Determination of morphine in the hair of heroin addicts by high performance liquid chromatography with fluorimetric detection. *J. Anal. Toxicol.* 1986;10:158-161.
27. Ishiyama, I., T. Nagai, S. Toshida. Detection of basic drugs (methamphetamine, antidepressants, and nicotine) from human hair. *J. Forensic Sci.* 1983;28:380-385.
28. Valente, D., M. Cussini, M. Pigliapochi, and G. Vansetti. Hair as the sample in assessing morphine and cocaine addiction. *Clin. Chem.* 1981;27:1,952-1,953.
29. Bailey, D. N. Drug screening in an unconventional matrix: Hair analysis. *JAMA* 1989;262:3,331.

5
The Medical Review Officer Function

Robert Swotinsky, M.D., M.P.H.

The federal approach to workplace drug testing places a physician, functioning in a role called the Medical Review Officer (MRO), between the laboratory and the employer. The MRO role was created to avoid characterizing as drug abusers people who have tested positive for reasons other than illicit drug use. Without such protection, drug testing would likely have been rejected by the courts and the public.

The MRO role first appeared in the 1988 Department of Health and Human Services (DHHS) "Mandatory Guidelines for Federal Workplace Drug Testing Programs."[1] The DHHS Mandatory Guidelines—also called the NIDA (National Institute on Drug Abuse) Guidelines—define the MRO as "a licensed physician . . . who has knowledge of substance abuse disorders and has appropriate medical training to interpret and evaluate an individual's positive test result together with his or her medical history and any other relevant biomedical information." The 1989 Department of Transportation (DOT) Procedures further require that the MRO not be responsible for, or be subject to supervision by, the laboratory's drug testing or quality control operations.[2] This latter requirement seeks to avoid conflict of interest.

The prerequisites for MROs are not well defined. The MRO need not take a course, pass an exam, belong to any organization, or be board-certified in any particular specialty. Some have argued that non-physicians should be allowed to serve as MROs. Others have argued for more rigorous requirements, for example, formal certification and continuing education programs. NIDA has considered this issue, and in 1990 stated that the MRO role should continue to be limited to physicians, and that MROs should be required to undergo continuing education in MRO functions.[3]

Legislation introduced in 1991 by Senators Hatch and Boren would establish standards for private-sector drug testing programs. The Hatch-Boren bill defines the Medical Review Officer as a licensed physician, registered nurse, or other trained individual.[4] The bill departs from the federal procedures in this and other areas in order to provide flexibility to employers.

It is desirable for the MRO to have experience in occupational medicine, since that experience sensitizes the physician to workplace requirements, confidentiality issues, and employee-management interactions. The MRO should also have knowledge of drug abuse and familiarity with the literature and procedures of drug testing. Nevertheless, for many MROs, the forensic, highly regulated drug testing procedures are an atypical and relatively small part of their practice.

This chapter presents a standard operating procedure (SOP) for the MRO function as established by the federal model of workplace drug testing. These procedures must be followed in drug testing programs in the federal workplace and in much of the transportation and nuclear power industries. Additional procedures and criteria may be established in the employer's policy, which the MRO should review before initiating services. Nonregulated drug testing programs may use procedures different from those required in federally mandated programs. It is desirable to follow the federal procedures because they help make drug testing reliable and confidential, and are the standard of care for physicians serving as MROs. Many of the procedures in this chapter also apply to the handling of alcohol test results, which is specifically addressed after the SOP. The MRO's role in rehabilitation and in return-to-work determinations is also discussed in this chapter.

STANDARD OPERATING PROCEDURE

Step 1. Receive the Chain of Custody Form

A chain of custody form, also known as a COC form, or custody and control form, establishes an audit trail so that the handling of a specimen can be reconstructed from collection through analysis, storage, and the final disposition of the specimen. At a minimum, the COC form identifies the specimen and records, in a tabular block, the name of each person who handled it. An external COC form tracks each specimen from the collection site to the laboratory. The laboratory maintains an internal chain of custody to track each specimen as it goes through accessioning, analysis, and storage. The employer and MRO can obtain the laboratory's internal chain of custody upon request.

DOT-mandated programs must use external COC forms consistent with the format illustrated in the DOT Procedures (see Appendix C). These forms have multiple copies as follows:

Copy 1: Laboratory copy
Copy 2: MRO copy, routed through the laboratory to the MRO
Copy 3: MRO copy, sent directly from the collection site to the MRO
Copy 4: Donor's copy
Copy 5: Collection site copy
Copy 6: Employer's copy, sent directly from the collection site to the employer
Copy 7: Laboratory copy for a split specimen

The copies are distributed as shown in Figure 5-1.

The copies have different levels of information. For example, the donor's name and phone number appear only on the donor and MRO copies. Reporting delays can occur when the MRO and the collection site are in different locations. The collection site may mail copy 3 to the MRO, or may mistakenly retain it; in either case, the MRO would usually receive the laboratory result prior to copy 3, which has the donor's name and telephone number. Until copy 3 is received, the MRO may be unable to identify or get in touch with the donor. To prevent this potential delay, the collection site can send copy 3 to the MRO by overnight courier or by facsimile. A less ideal alternative is to ask the employer for the donor's name and telephone number corresponding to the ID number on copy 2. This request should, if possible, be posed to an employer's representative who will keep confidential the names of donors with laboratory-positive results that are undergoing review. DOT-mandated programs should not allow copy 3 to be routed to the MRO through the laboratory or the employer, as this violates the DOT Procedures. Nevertheless, an accidental misrouting of copy 3 to the laboratory does not invalidate the test; the laboratory is unlikely to treat a specimen differently based on the examinee's name.

DOT requires that chain of custody forms be received and reviewed at the MRO's office before the results are reported. Forms corresponding to laboratory-positive results undergo a detailed review; forms corresponding to laboratory-negative results undergo a less substantial review. The NIDA Guidelines and the Nuclear Regulatory Commission (NRC) regulations also require that the MRO receive a copy of the chain of custody form. Chain of custody forms are the primary record of the collection procedures, and should be received and reviewed before reporting results. Review of the forms is especially important for positive results, which should undergo the highest level of scrutiny. Receiving and reviewing chain of custody

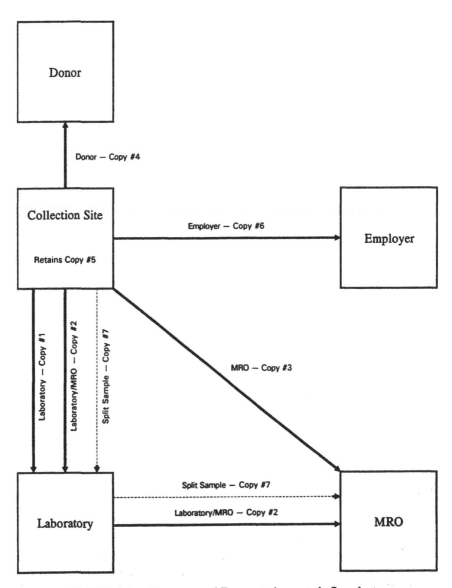

FIGURE 5-1. Department of Transportation custody flow chart.

forms increases turnaround time, but allows the MRO to ensure that the COCs are adequately completed. NIDA has instructed federal agencies to ensure that their MROs receive and review chain of custody forms prior to reporting results.[5] In nonregulated programs, the extent of the MRO's review of COC forms is ultimately decided by both the employer and the MRO, based on costs, potential reporting delays, and the perceived benefit of the MRO's added involvement.

Step 2. Receive Results from the Laboratory

The laboratory must transmit all results directly to the MRO's office; they cannot pass through the employer, through a consortium administrator, or through any other third party. Laboratory reports are printed in a variety of formats. Laboratory results and detected drugs, if any, are also identified on copy 2 of the COC form when a DOT form is used. Laboratory reports may be transmitted in print or by secure electronic transmission (teleprinter, facsimile, or computer-to-computer). Verbal (telephone) reports are prohibited because they are more easily misunderstood and the transmission is less secure. Drug test results are well suited for computer-to-computer transmission because their format is relatively consistent (in comparison with a variety of other medical tests). Some MROs access and manipulate electronic results, accomplishing quicker turnaround, less paperwork, and lower costs.

The NIDA Guidelines and NRC regulations[6] require batch reporting. Reporting of both positive and negative results awaits completion of the batch, which consists of all specimens submitted at the same time to the laboratory. Batch reporting is designed to make specimens that undergo confirmation testing indistinguishable from those that do not. DOT does not require batch reporting. There is a growing consensus that reporting delays caused by batch reporting outweigh the potential benefit.[7]

The laboratory reports which, if any, drugs or metabolites are detected at or above the cutoff levels. (Tests in which drug or metabolite levels meet or exceed the cutoffs are referred to in this chapter as "laboratory-positive.") The laboratory will, upon request by the MRO, report quantitative results (i.e., levels) for analytes in specific cases. The laboratory will, upon written request by the MRO, routinely report opiate levels of all opiate-positive results. (Quantitative opiate results can help distinguish codeine from morphine use.) The laboratory cannot routinely report levels, with the exception of opiates, for tests conducted in accordance with NIDA's procedures,[8] but may routinely report quantitative results for nonregulated tests upon request.

Some employers would prefer to receive negative results directly from the laboratory, thereby reducing the reporting time for negative results and eliminating the expense of the MRO. Direct transmission of results to employers raises confidentiality concerns. If the employer receives some negative reports from the laboratory and others from the MRO, the employer can recognize that the MRO's negative reports are laboratory-positive results that the MRO made negative. (This chapter refers to such tests as "MRO negatives.") People with MRO negatives might then (unfairly) be subject to adverse employment decisions, unwarranted labeling, or job discrimination. NIDA and DOT have taken the position that this confidentiality issue outweighs the possible added time and expense of sending laboratory-negative results to the MRO. However, DOT has been under particular pressure to reconsider this issue.

In 1990, DOT proposed an amendment to its procedures that would affect the routing of laboratory-negative results. This proposed amendment contained two alternate ways of handling laboratory-negative results, both designed to maintain confidentiality while avoiding the perceived shortcomings of sending negative results to the MRO.

- Proposed Alternative 1 (Figure 5-2): The laboratory transmits all results to a designated employer representative. The employer representative sends laboratory-positive results to the MRO and laboratory-negative results to company officials. The MRO reports MRO-negative tests to the employer representative, who reports them to company officials. The MRO reports verified positives to the company directly or through the employer representative. The employer representative does not supervise the donor and is not able to take adverse action against the donor. A figurative "bubble" around the employer representative prohibits others from learning of laboratory-positive results that have not been verified by the MRO. (Critics have said that those with confidence in this arrangement are themselves living in a bubble.)

- Proposed Alternative 2 (Figure 5-3): The laboratory transmits negative results directly to the company and positive results to the MRO. The MRO reports verified positive tests directly to the company, and reports MRO-negative tests to the laboratory, which reports them to the employer. In this way, the employer receives all negative reports from one source. This alternative is intended for use in small companies where it is impractical to designate an employer representative.

These alternatives have generated strong opinions. MROs can profit from handling laboratory-negative results, and advocate their role in protecting confidentiality and in monitoring the administrative and technical

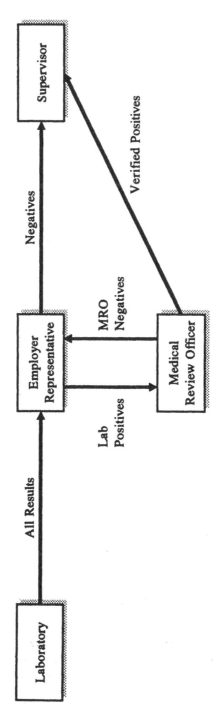

FIGURE 5-2. Proposed Alternative 1.

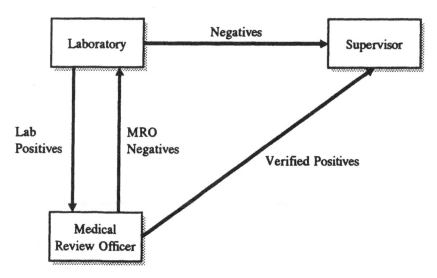

FIGURE 5–3. Proposed Alternative 2.

adequacy of the testing. Some laboratories object to the administrative responsibility that alternative 2 assigns to them. Given the divided opinions, the complexity of implementing changes, and the confidentiality issues, DOT is unlikely to adopt either alternative.

In some nonregulated programs only laboratory-positive results are transmitted to the MRO. This should be discouraged because it identifies MRO-negative results to the employer and significantly constrains the MRO's ability to provide quality assurance, for example, by independent review of laboratory-negative COC forms.

In addition to getting donors' results, the MRO usually also receives analytic results of performance testing specimens. These specimens are submitted at collection sites by the employer or the employer's designate, sometimes the MRO. The MRO must know how to identify and handle these results.

Step 3. Declare Tests with Significant Procedural or Technical Errors as Invalid

The laboratory checks that each COC form is complete and matches the ID on the accompanying specimen container. The laboratory also checks whether the specimen and package seals are intact, whether the volume is sufficient, and whether the specimen's condition suggests adulteration. If the packaging, COC form, or specimen condition is significantly defective, the laboratory notes the problem on the COC form and asks the employer—or the MRO, who often acts in the employer's stead—whether or not to analyze the specimen. The employer or the MRO then have the option of canceling the analysis and declaring the test invalid. An invalid test, or no-test, is neither negative or positive.

The DOT Procedures require that COC forms corresponding to negative results undergo an administrative review at the MRO's office. This review can be conducted on any copy of the COC form. DOT recommends that the administrative review, at a minimum, verify that the donor ID on the COC form matches the donor ID on the laboratory report.[9] DOT requires the MRO to personally review copies 2 and 3 of COC forms that correspond to positive results, to also ensure that the collector, donor, and laboratory signatures are present, and that the chain of custody blocks are intact.[9]

There is a growing consensus on which particular errors are significant enough to cause invalidation. The employer usually depends on the laboratory, sometimes with the MRO's oversight, to determine which tests are invalid. But when there is disagreement, the final decision is the employer's.

DOT's recommended criteria for invalidating DOT tests are presented in Figure 5-4. DOT distinguishes those errors that are not correctable from those that are. NIDA recommends that laboratories retain invalid specimens for at least 5 working days, which allows time for the employer or MRO to take corrective action. A correction can be made by submitting to the laboratory or MRO a corrected form and a written statement (affidavit) by the individual who made the error. Figure 5-5 is an example of a corrective affidavit form.

A complete chain of custody block refers to a DOT-type COC block that, at a minimum, has the collector's signature twice and the courier's name once, as in Figure 5-6. Non-DOT chain of custody blocks may have different formats (e.g., may include the donor's signature, or have only one signature column). By analogy with DOT's requirements, these blocks should be considered complete only if the courier name, collector's signature, and donor's signature (or documentation that the donor refused to sign) appear.

The employer's policy may impose additional requirements that define when a test is unacceptable. If the employer has no policy or procedures, the MRO should follow widely accepted, reliable guidelines, such as those of DOT.

The laboratory performs tests—typically color, smell, pH, specific gravity, creatinine concentration, and/or osmolality—on each specimen to determine if it is adulterated or diluted. NRC specifically allows additional processing (e.g., testing for additional drugs or lower cutoff levels) of specimens in which adulteration or dilution is suspected. An obviously adulterated specimen (e.g., orange juice, dog urine) should be considered a refusal

---

Chain of Custody Form: Correctable Flaws

1. Donor ID number is omitted on form unless the donor's refusal to provide the ID number is noted on the form.
2. An incomplete chain of custody block (Minimum: two signatures, shipping entry, dates).
3. Collector's signature is omitted from the certification statement.
4. Donor's signature is omitted from the certification statement unless the form elsewhere indicates that the donor refused to sign it.
5. Certifying scientist's signature is omitted on positive results.

Chain of Custody/Specimen Condition: Fatal Flaws

1. Specimen ID number is omitted from specimen bottle, or does not match the specimen ID on the chain of custody form.
2. Specimen volume below 30 ml. If the volume is between 27 and 30 ml, the specimen may be accepted if the laboratory can ensure that sufficient volume will be available for storage and any necessary reanalyses for quality control or reconfirmation of results.
3. Specimen bottle seal is broken (e.g., specimen leaks) or shows evidence of tampering.
4. Specimen shows obvious adulteration (i.e., color, foreign objects, unusual odor).*

[*Note: If the collection site finds the specimen obviously adulterated or its temperature unacceptable, DOT requires the site to collect a second, witnessed specimen. Both specimens are then valid and undergo analysis.]

FIGURE 5-4. Recommended test invalidation criteria.
Source: Adapted from U.S. Department of Transportation.[10]

to undergo testing; the specimen should be retained in case of legal challenges. Abnormal creatinine or specific gravity test results should not be cause for invalidating a test. A dilute specimen at a previous test, according to DOT and NRC, is valid, and is grounds for collecting a subsequent specimen under direct observation. Sometimes the laboratory does not analyze a specimen that falls below dilution criteria, for example, per an employer's request. If a specimen is not analyzed because it is dilute, the MRO should report the result as invalid and dilute.

Dilute urine specimens can indicate adulteration, water loading, diuretic use, or medical conditions such as diabetes, kidney disease, and psycho-

I, _____, underwent a drug test specimen collection
   Donor's Name

at _____. On _____,
   Collection Site Name and Address   Collection Date

I omitted my signature from the certification statement on the drug test custody and control form _____. I am submitting this affi-
                                        ID Number
davit to correct and amend the form.

_____ (Print)

_____ (Signature)

_____ (Date)

(This section is optional)

Sworn and subscribed before me
this _____ day of _____, 19___.

_____ (SEAL)

Notary public of: _____

My commission expires: _____

(Send the completed affidavit to the laboratory or the MRO.)

FIGURE 5-5. Affidavit for omitted donor certification statement signature.

PURPOSE OF CHANGE	RELEASED BY	RECEIVED BY	DATE
Provide Specimen for Testing	- DONOR -	*(Collector)*	*(Date)*
	(Collector)	*(Courier)*	*(Date)*

FIGURE 5-6. Essential entries on the DOT chain of custody block. (Entries in italics are completed at the collection site.)

genic polydipsia. Water loading can be intentional or inadvertent, for example, it may be induced by fluids given at the collection site in order to overcome a "shy bladder."[11] Intentional water loading is unlikely if the specimen was collected on short notice, for example, a post-accident, random, or reasonable cause test. In some non-DOT programs, employers adopt a policy of declaring dilute specimens invalid or requiring that they be followed promptly by witnessed retests. Given the many possible causes of dilute specimens, these actions may be unwarranted, particularly when testing a population with a low rate of illicit drug use. The MRO should notify the employer of dilute specimens, defined as specific gravity below 1.003 and/or* creatinine concentration below 0.2g/L. This information should be kept separate from any written report of the test's determination that is sent to the employer, because it can make the determination appear less sound.

Step 4. Review and Interpret Laboratory-Positive Results

The MRO reviews each valid, laboratory-positive result. If the MRO finds no medically legitimate reason for the result, it is reported to the employer as "verified positive." If a medically legitimate reason is found, it is reported as "negative." If the MRO believes the test has a significant procedural or technical error, it is reported as "invalid."

Laboratory-negative drug tests can be handled administratively, that is, checked for completeness and adulteration criteria, and then reported to the employer. The MRO's staff can handle negative results without the MRO's direct involvement. The MRO's staff can also initiate contact with donors who test laboratory-positive. The DOT Procedures allow medically licensed staff (a licensed nurse, medical technician, etc.) to gather information pertaining to positive results. If the MRO and donor do not speak the same language, a translator can assist.

---

*The DOT Procedures state "and"; the NRC regulations state "or."

Step 4-a. Contact the Donor
The MRO must give each examinee with a valid laboratory-positive result an opportunity to discuss the result in person or by telephone. This is true even if the donor tells the MRO's staff he or she used the identified drug illegitimately. The employer can help determine procedures for contacting donors. Face-to-face encounters are optimal, but telephone discussions are more expedient and are common practice, especially when the MRO and donor are located far apart. The donor's telephone number is on copy 3 of DOT chain of custody forms. The donor's telephone number—home, work, or both—may be on other chain of custody formats, too, or may be on medical records or other documents that may be accessible to the MRO. Calling the home number is preferable. The work number may be more convenient, but a call at work may prompt a confrontation between the donor and management that would not occur if the donor was notified at home and was more composed. Calls at work are also prone to discovery by other employees.

The MRO's review can be initiated in DOT-mandated testing immediately upon receipt of the laboratory-positive result. The MRO cannot verify the result as positive until both copies 2 and 3 of the COC form are received and reviewed. By reviewing DOT drug test results only after both COC copies are received and reviewed, the MRO can avoid potentially unnecessary action on invalid tests.

The character and content of each attempted and actual communication should be documented. Figure 5-7 demonstrates a format for recording drug test reviews. A punchlist or similar questionnaire or chart may be added to the MRO's record of each laboratory-positive result. Comments may include answering machine recordings, messages with friends or relatives, and disconnected or wrong numbers. Thorough documentation helps establish that the donor has been given an opportunity to discuss the results.

If someone other than the donor answers the phone, the MRO (or MRO staff assistant) should leave a brief message requesting that the donor return the call. If asked, the MRO might explain that he/she is a physician who works for the donor's employer, and that the call pertains to a recent examination. The MRO should protect the donor's privacy and not reveal the specific reason for the call.

Under the DOT Procedures, when the MRO cannot reach a donor within a reasonable period, the MRO may ask a designated employer's representative for assistance. If the donor is self-employed without a staff—for example, an independent truck driver—the MRO should send a certified letter to the donor, return receipt requested, asking the donor to respond. What constitutes a "reasonable period" depends on the MRO's judgment, which may vary with the employer's needs. If the employer's representative can-

FIGURE 5-7. The Medical Review Officer punchlist.

Donor Name: _____ I.D. No. _____

✓ Are there significant packaging or chain of custody errors? <u>Yes</u> / <u>No</u> IF "YES," THE TEST IS "INVALID."

NOTE: Provide the donor with an opportunity to discuss positive test result(s). Document all attempts. If unable to contact the donor reasonably soon, ask the employer's representative to assist.

<u>DATE</u>	<u>TIME</u>	TELEPHONE NUMBER	CONTACT? (Y/N)	<u>COMMENTS</u>
___	___	(___) ___-_____	Y/N	_____
___	___	(___) ___-_____	Y/N	_____
___	___	(___) ___-_____	Y/N	_____
___	___	(___) ___-_____	Y/N	_____
___	___	(___) ___-_____	Y/N	_____
___	___	(___) ___-_____	Y/N	_____

(continue on the back of this sheet)

✓ Identify yourself as a physician and Medical Review Officer.

✓ Confirm the individual's identity, e.g. Read 5 digits of their soc. sec. #, and ask for the other 4 digits.

✓ Tell the donor that medical information that affects fitness for duty may be shared with the employer.

✓ Inform the donor of the test results.

✓ Seek potential sources of the drugs, e.g. legitimate medications, invasive procedures, poppy seeds, etc.

✓ Request the donor's assistance in confirming that prescription(s) are legitimate.

✓ If positive only for opiates, and not known positive for 6-MAM, look for clinical evidence of opiate abuse.

✓ If positive for methamphetamine, determine isomeric composition ($d = R_x$, $l = $ OTC)

✓ Consult with laboratory officials and/or other experts, as needed.

✓ Request, as needed:
- ☐ Quantitative results
- ☐ Reanalysis of the specimen, e.g., per donor's request.
- ☐ Analysis of the split specimen (if collected)

DATE	TIME	TELEPHONE NUMBER	LAB CONTACT PERSON	COMMENTS
		( ) -		

✓ Based on your review, make one of the following determinations:

____ Negative   ____ Verified Positive   ____ Invalid

Report results to the appropriate official.   REPORTED BY: ____

MRO SIGNATURE: ____   REPORTED TO: ____   ON: __/__/__

---

not contact the donor, the employer may place the donor on temporary medically unqualified status or medical leave; the test remains unresolved.

After contact from the employer's representative, the donor has 5 days in which to contact the MRO. After 5 days, the MRO may report the result as verified positive, even if the MRO has not reached the donor. The MRO may also report the result as verified positive if the donor expressly refuses to discuss the test. These are the only two situations in which an MRO may report the test as positive without having spoken with the donor. Some employers prefer that the MRO denote these as "non-contact-verified positives." The Federal Highway Administration (FHWA) requires that the MRO provide the employer with documentation of efforts made to contact the driver in such cases.[12] The DOT Procedures require the MRO to reopen the case and re-review the results if the donor later demonstrates that extenuating circumstances (for example, a serious illness or injury) prevented contact.

Step 4-b. Confirm the Donor's Identity

John Smith Jr. rather than Sr. (or vice versa), or someone else altogether, may answer the telephone. An attempt should be made to establish that the person who is contacted is, in fact, the donor. For example, the MRO can read the first part of the donor's ID (employee ID or social security number), and ask for the latter half. This technique also tells the donors that the MRO knows something about them, and thereby helps establish the MRO's legitimacy. Alternatively, the MRO can ask for the donor's birthdate, employer, and so forth.

Step 4-c. Optional: Tell the Donor That Fitness-for-Duty Information May Be Disclosed

The DOT Procedures allow the MRO to convey medical fitness-for-duty information—other than the drug test results—to the employer, a DOT or other federal safety agency, or a physician who examines the donor for medical fitness under DOT regulations. For the MRO to disclose such information, DOT requires that the MRO tell the donor, before obtaining the information, that it may be disclosed. (This process is analogous to reading the Miranda rights: "Anything you say can and will be used against you. . .") Furthermore, the MRO can disclose the information only if:

1. An applicable DOT regulation permits or requires such disclosure;
2. In the MRO's reasonable medical judgment, the information could result in the donor being determined to be medically unqualified under an applicable DOT agency rule; or,
3. In the MRO's reasonable medical judgment, where there is no applica-

ble DOT agency rule establishing physical qualification standards, the information indicates that continued performance by the donor of his or her safety-sensitive function could pose a significant safety risk.

Compliance with the first and second items above requires knowledge of DOT agency rules (including those for aviation, motor carrier, railroad, maritime, mass transit, and pipeline industries) and DOT medical standards (e.g., those of FHWA[13] and FAA[14]). Compliance with the third item above may be even more complex, given that "reasonable medical judgment" leaves much to discretion, and can be a difficult standard to defend, for example, in a negligence lawsuit.

In practice, disclosure is more of an obligation than an option. Disclosure of fitness-for-duty information can potentially prevent accidents. And, if the MRO chooses not to disclose, he or she may be held liable in the event of a subsequent accident.

Although the MRO can disclose information incidental to the drug test result, the donor comes to the MRO only because of the drug test, not for an assessment of his/her medical condition. The MRO should therefore attempt to restrict the focus of each review to the specific drug(s) for which the donor tested positive.

Conditions that might prompt disclosure include the legitimate use of sedatives or stimulants, and the presence of medical conditions that might lead to sudden incapacitation. The information that is disclosed should be limited to appropriate restrictions or recommendations for follow-up. The employer should not be given diagnoses or details of a specific nature.[15]

Step 4-d. Inform the Donor of the Test Result
The donor should be told which substances were detected. This helps the donor respond with potential legitimate sources for those substances. The donor's initial reaction can be telling, so the MRO should personally make this contact, rather than delegating it to a staff member.

Step 4-e. Seek Potential Sources of the Drug(s)
The MRO seeks potential legitimate explanations for positive results. These explanations can be:

1. Legitimately prescribed or dispensed controlled medication(s);
2. Over-the-counter medication(s);
3. Poppy seed ingestion, which can cause opiate-positive results;
4. Analytic or reporting errors by the laboratory.

Legitimately prescribed or dispensed medications are the most common explanations. Poppy seeds are a less frequent explanation; opiate-positive results are more often due to prescription medications. Analytic error is suspected in only the most unusual of situations. Properly run, immunoassay screening followed by gas chromatography/mass spectrometry (GC/MS) testing virtually eliminates the possibility of false-positive results (finding drugs that are not there) or misidentification (findings drugs that are there but calling them the wrong drugs).[16,17]

The MRO can ask about, and the donor may admit to, illegitimate use of the identified drug. A donor admission to the MRO of illegitimate drug use equates with a verified positive determination.

The MRO seeks from the donor potential legitimate source(s) of the positive result(s). The donor may not recall or recognize having used the drug. This is particularly true of cocaine, which is used as a topical anesthetic. If the donor tests positive for cocaine, the MRO should ask about potential sources, such as recent nose, eye, or throat surgery, dental procedures (this is outdated and rare), bronchoscopy, repair of lacerations, and prescription eye drops.[18,19] If the donor tests positive for a drug that is found in medications, the MRO should ask the donor to identify medications taken before the collection. The MRO must consider the half-life of each drug and its metabolites, which ranges from a few days for cocaine[20] to 4-6 weeks or more for marijuana.[21,22] Table 5-1 presents some legitimate sources for marijuana, cocaine, opiates (morphine and codeine), and amphetamines (amphetamine and methamphetamine). Phencyclidine (PCP) has no medical applications and is not found in nature; its use is always illegitimate.

Step 4-f. Verify Sources of the Drug(s)

The federal regulations require that the MRO review medical records provided by the donor to substantiate prescription medication(s) that could have caused the result. Beyond this, the verification process is left to the MRO's judgment.

Donors sometimes list, usually at the collection site, recently taken medications. In DOT programs, these lists are "memory joggers" for the donor's use, only. In non-DOT programs, the MRO often receives these lists. Self-reported lists may be incomplete or inaccurate. Donors should not be penalized for neglecting to record medications, nor should self-reported medications be accepted without corroboration.

Corroboration is best obtained directly from the pharmacist or the prescribing medical provider. Less rigorous evidence (e.g., a prescription label, the donor's assurance) runs a risk of deception. Medical providers, with good reason, may be reluctant to discuss their patients with an unfamiliar person, particularly in the context of drug testing. In anticipation of this,

TABLE 5-1 Medications That Can Cause Laboratory-Positive Drug Test Results

	Nonprescripton Medications	Prescription Medications

Amphetamines	l-methamphetamine
	Vicks Inhaler®
		amphetamine
		  Biphetamine Capsules®
		methamphetamine
		  Desoxyn Gradumet Tablets®
		Selegiline (Eldepryl®), benzphetamine (Didrex®), and deprenyl[23] metabolize to both amphetamine and methamphetamine

Cocaine	(none) (Health Inca Tea contains cocaine,[24] but should no longer be available in the United States.)	cocaine, used as a vasoconstrictive anesthetic, e.g., in otolaryngology, ophthalmology, and dentistry TAC[25] (tetracaine, adrenaline, and cocaine mixture used in some emergency rooms for suturing) Brompton's mixture (contains an opiate—usually morphine or heroin— and cocaine and/or a phenothiazine. Used for pain control of the terminally ill.)
Marijuana	(none)	dronabinol Marinol Capsules®

Opiates	Acetaminophen with codeine is available in Canada and some other foreign countries without a prescription.	codeine
		  Acetaminophen with Codeine®
		  Actifed with Codeine Cough Syrup®
		  Ambenyl Cough Syrup®
		  Aspirin with Codeine®
		  Broncholate CS®
		  Calcidrine Syrup®
		  Capital with Codeine Suspension®
		  Codalan Tablets®
		  Codimal PH®
		  Dimetane-DC Cough Syrup®
		  Empirin with Codeine®
		  Fiorinal with Codeine®
		  IoTuss Liquid®
		  Isoclor Expectorant®
		  Naldecon CX Liquid®
		  Novahistine DH®
		  Nucofed
		  Pediacof Cough Syrup®
		  Phenaphen with Codeine®
		  Phenergan with Codeine®

(continued)

TABLE 5-1 Medications That Can Cause Laboratory-Positive Drug Test Results (Continued)

	Nonprescription Medications	Prescription Medications

Opiates (cont.)		codeine (cont.)
		    Poly-Histine CS®
		    Promethazine with Codeine®
		    Robitussin A-C®
		    Robitussin DAC®
		    Soma with Codeine®
		    Terpin Hydrate with Codeine®
		    Triaminic Expectorant with Codeine®
		    Triprolidine Pseudoephedrine with Codeine®
		    Tussar-2®
		    Tussi-Organidin Liquid®
		    Tussirex Syrup®
		    Tussirex Sugar-Free®
		    Tylenol with Codeine (#1, 2, 3, or 4)®

	morphine[a]
	    Donnagel-PG®
	    Parepectolin Suspension®
	    Paregoric®
		morphine
		    Astramorph/PF Injection®
		    Duramorph®
		    MS Contin Tablets®
		    MSIR®
		    OMS Concentrate®
		    RMS Suppositories®
		    Roxanol®
		    Roxanol Suppositories®

	Poppy seeds contain morphine and smaller amounts of codeine	
Phencyclidine	(none)	(none)

[a] Certain Schedule V substances are available without a prescription. The pharmacy must maintain a Schedule V record book with the name and address of each purchaser, the name and quantity of the controlled substance purchased, and the date of sale.[26]

the donor should be asked to authorize release of the information to the MRO. The donor can be asked to directly contact the medical provider to authorize the release; if so, the donor should be asked to notify the MRO once this has been accomplished. Alternatively, the donor can provide the MRO with a written release that the MRO can then forward to the medical provider. The donor should be given reasonable deadlines for providing the release. If the donor does not cooperate with the review, the MRO can inform the employer, who can place the donor on administrative or medical leave pending resolution of the drug test. Alternatively, the MRO can verify the result, if nonopiate, as positive.

The medical provider can provide information in person, in writing, or by telephone. If a letter is received, especially if it is not on office letterhead, a follow-up telephone call can help confirm that the letter is authentic. If corroboration is sought by telephone, the MRO should get the medical provider's telephone number from the donor, telephone directory, or directory assistance and should then initiate, rather than receive, the call. These strategies decrease, but do not eliminate, the possibility of deception.

If the MRO determines that the medication was improperly used (e.g., too much was taken) or was improperly prescribed, the result may be verified positive even when a physician's prescription exists. In this case, the determination should be firmly substantiated; the donor is especially likely to challenge it.

Step 4-g. Do Not Verify Opiate-Positive Results Without Clinical Evidence of Opiate Abuse (Unless 6-MAM Is Detected)

Poppy seeds contain morphine and codeine, and can produce urine opiate levels that exceed 300 ng/ml. Poppy seeds are the only legitimate nonpharmaceutical sources of positive drug test results. Many foods contain poppy seeds, for example, poppy seed bagels, poppy seed bread, veal with poppy seeds, poppy seed sauce, poppy seed torte, and poppy seed cake. The MRO should not base the test's determination on the donor's recall of what foods he or she ate, or on the donor's knowledge of whether those foods contained poppy seeds. The donor may not remember or recognize sources of poppy seeds. The donor may falsely claim poppy seed ingestion.

Poppy seeds contain more morphine than codeine, and usually produce urine codeine levels below 300 ng/ml and morphine-to-codeine ratios that exceed 2:1.[27] Codeine levels and morphine-to-codeine ratios are not, however, infallible ways to distinguish codeine from morphine consumption.[28] The federal regulations allow the MRO to verify the result as positive only if 6-monoacetyl morphine (6-MAM, a heroin metabolite) is detected or if the MRO finds clinical evidence, in addition to the urine test, of opiate abuse. The burden of proof shifts to the MRO: The donor need not demonstrate legitimate drug use; instead, the MRO must demonstrate clinical evidence of opiate abuse. NRC extends this approach to drugs that are commonly prescribed or that are found in over-the-counter preparations: The MRO must determine if there is clinical evidence, in addition to the urine test, of abuse corresponding to such drugs in order to verify a laboratory-positive NRC drug test result. NRC's requirement is rather restrictive and, in the author's opinion, need not be applied to non-NRC testing for non-opiate drugs.

6-MAM is specific for heroin use, which is clearly illegitimate. 6-MAM is detectable in the urine for only a few hours following heroin use[28] and is unstable. Its presence confirms heroin use, but its absence does not disprove heroin use. The analysis is expensive. The MRO may need to have the specimen transported to a different laboratory; few laboratories perform 6-MAM analyses. Because of these limitations, it is appropriate to be selective in performing 6-MAM analyses of opiate-positive specimens (except in NRC-mandated testing, where all morphine-positive specimens must undergo 6-MAM analyses[30]). Some laboratories automatically perform 6-MAM analyses of any specimen with a morphine concentration that meets or exceeds a threshold, for example, 2,000–5,000 ng/mL.

Without detection of 6-MAM or clinical evidence, the test cannot be a verified positive. Clinical evidence of abuse can come from the MRO's discussion with the donor—for example, the donor admits to abuse of the drug that was detected—or from a medical examination of the donor by the MRO or another clinician with appropriate expertise. The clinical evidence must be of recent use, for example, needle tracks suggest heroin use, but old needle track scars do not explain a recent opiate-positive result. Observations by staff at the collection site or by supervisors at the work force are less reliable, but may be useful if supplemented with other clinical evidence.

The donor may refuse a medical evaluation, thereby removing this potential source of clinical evidence. The employer, not the MRO, can insist that the donor submit to the evaluation.

Given these constraints, and given that the rates of heroin use and illicit codeine or morphine use are relatively low, a small proportion of opiate-positive drug tests are verified positive by the MRO. Why do the federal regulations include opiates among the required tests? Opiates are included in large part because the federal procedures evolved from the military programs that screened for heroin and marijuana use among servicemen returning from Vietnam.

Step 4-h. Consult with Laboratory Officials and/or Other Experts, as Needed

If the accuracy or validity of a positive test result is in doubt, the MRO should review the laboratory records to determine whether the required procedures were followed. This may require collaboration with the laboratory director, the analysts, and expert consultants. Such inquiries can have important consequences, for example, in 1990, an MRO's inquiry helped prompt an investigation that uncovered methamphethamine analytic errors at several NIDA-certified laboratories.

Step 4-i. If Positive for Methamphetamine, Determine Isomeric Form

Methamphetamine has two stereoisomers, each with distinct pharmacologic properties: The *d* (dextro) form is a strong stimulant and has a high abuse potential. The *l* (levo) form is weaker, and is the active ingredient in Vicks Inhaler®, an over-the-counter decongestant. Illegally manufactured methamphethamine contains the *d* form but may also contain significant amounts of the *l* form. If the MRO suspects a positive methamphetamine result may be due to Vicks Inhaler, the MRO should clarify whether the laboratory found the *d* or *l* isomer. Some laboratories routinely differentiate between these isomers and only report the *d* form. If not, the MRO should ask the laboratory to determine the methamphetamine's isomeric form. Vicks Inhaler use may be associated with low concentrations of the *d* isomer. Low concentrations of the *d* isomer produced during sample preparation may be measured because of analytic cross-reactivity. NIDA has recommended that identification of *l*-methamphetamine at more than 80 percent is consistent with the use of Vicks Inhaler and should, therefore, be interpreted as negative.[31]

Step 4-j. Request Quantitative Results, Reanalysis of the Specimen, and/or Analysis of the Split Specimen (If Collected), as Needed

If the MRO suspects excessive use, or if the result's validity is in question, the concentration may be useful. The MRO can request from the laboratory the drug or metabolite concentrations of specific specimens. The MRO should first verify that the concentrations meet or exceed applicable cutoffs. High concentrations raise a suspicion of abuse, but may also represent a recent appropriate dose, or an intentional or accidental inappropriately high dose. Urine levels depend on pharmacokinetics, urine flow, urine pH, and other factors. Because of wide individual variations in these parameters, urine levels should be interpreted only within broad limits.

Under federal procedures, the MRO has the sole authority to authorize reanalysis of the specimen. A reanalysis may be indicated if the accuracy of a laboratory-positive result is in question. The DOT Procedures require that the MRO authorize reanalysis if requested in writing by the donor within 72 hours after being notified of the result. DOT's aviation,[32] pipeline,[33] and railroad[34] regulations extend this period to 60 days after receipt of the result by the MRO. NRC requires that the donor's request be "timely." The employer's policy may also address reanalyses, including designation of the responsible party—donor or employer—that pays for reanalyses.

Some employers authorize a split-sample collection procedure, in which the donor provides two specimens from a single void. The first specimen is analyzed, and the second is stored, usually at the laboratory, for analysis at a later time, should the first specimen test positive. The DOT Procedures require that the MRO authorize analysis of the second specimen if requested by the donor within 72 hours after being notified of the result. NRC's regulations require analysis of the second specimen at a different NIDA certified laboratory upon request by the donor.

Confirmation of initial results—that is, reanalyses and analyses of split specimens—are done by GC/MS testing for only the drugs or metabolites that were initially detected. This should be done at a different laboratory than the initial analysis, to provide a quality control check of the first laboratory. Drugs and metabolites may deteriorate with time. Because of analytic variability, results can fall above or below the cutoff when the true concentration is near the cutoff. For these reasons, a reanalysis or analysis of a split specimen need only demonstrate the presence of the drug or metabolite at any level, without regard to the cutoff, to confirm the initial positive result.

A reanalysis or analysis of a split specimen does not necessarily mean the initial result is suspect. If the MRO is confident of the initial result, the MRO can and should notify the employer of the outcome, without waiting for the second result. This entails a risk—unlikely but consequential—that the second result will not confirm the initial result, in which case the test's outcome must be changed to "negative."

Step 4-k. Determine Whether the Result Is Consistent with Legitimate Drug Use

Each test has one of three outcomes: negative, verified positive, or invalid. A laboratory-positive test has a "negative" outcome if:

1. The MRO determines that the donor's explanation is legitimate; or,
2. The result is opiate-positive (and not known positive for 6-MAM) and the MRO finds no clinical evidence of opiate use.

A laboratory-positive test has a "verified positive" outcome if:

1. The donor provides no explanation for a nonopiate result or provides an explanation that the MRO cannot corroborate; or,
2. The result is opiate-positive and the MRO finds clinical evidence of opiate use; or,
3. The result is opiate-positive and positive for 6-MAM.

The test's outcome is "invalid," without regard to the laboratory result, if a significant procedural or technical error occurred.

Certain federal agencies have additional conditions (see Figure 5-8) that must be met before a laboratory-positive result can be deemed "negative." The employer's policy may also help define when drug use is or is not acceptable. Beyond this, the MRO's determinations are based on professional judgment. The MRO is a physician, not a judge, who makes medical determinations of legitimacy, not legality. The MRO does not make employment decisions, and should not be influenced by potential repercussions of his/her determination. Perceptions of the donor's honesty also should not critically influence the MRO's decisions. The forensic nature of workplace drug testing demands facts, not perceptions. A consistent approach helps assure that donors are treated fairly and equally.

Reviews of positive tests should be expedited, for example, started on the day of receipt. NRC requires that, within 10 days of the presumptive positive screening test, the MRO complete the review and report the out-

---

Federal Highway Administration (FHWA).[36] The medication must have been prescribed or authorized by a licensed medical practitioner who was familiar with the donor's medical history and assigned duties.

Federal Railroad Administration (FRA).[37] FRA requires that:

1. The treating medical practitioner or a physician designated by the railroad has determined, with knowledge of the donor's medical history and assigned duties, that use of the medication and the prescribed or authorized dose does not pose a safety risk;
2. The medication was used at the dosage prescribed or authorized;
3. In the event that the donor is being treated by more than one medical practitioner, at least one treating practitioner has been informed of all medications authorized or prescribed and has determined that use of the medications does not pose a safety risk, *and* the donor has adhered to any restrictions imposed with respect to use of multiple medications.

Nuclear Regulatory Commission (NRC).[38] Use of the drug at the prescribed dosage must not reflect a "lack of reliability" or be likely to create on-the-job impairment.

FIGURE 5-8. Agency-specific conditions for acceptable medication use with respect to workplace drug testing.

come to the employer.[38] The Federal Railroad Administration (FRA) requires that the MRO complete the review within 10 working days of receipt of the laboratory result—barring delays caused by the donor—or else the test's outcome must be reported as "negative."[39] The employer, not the MRO, can authorize the donor to undergo another drug test. But, past and future drug test results have no bearing on the MRO's review of the drug test result. The MRO's conclusion—verified positive, negative, or invalid—is based on one specimen per test. An exception may be made only if the employer asks that the test be left unresolved, for example, if the donor is unreachable.

Complex issues include the following:

- A donor who tests positive may claim, truthfully, that he or she took a controlled substance that was prescribed for a spouse, relative, or acquaintance. Taking another person's medicine is not illegal under federal law. This kind of use, sometimes referred to as "spousal use", may appear medically indicated in selected cases. As with any review, the MRO must decide if the use was legitimate. DOT advises MROs that such results should be reported as verified positive.[9] The employer's policy may address this issue, for example, by designating that unauthorized use of a medication is illegitimate. The MRO's handling of such claims should be consistent. The MRO can, with the donor's permission, identify such results to the employer as "spousal positives," but should first verify that the spouse, child, or other acquaintance had a legitimate prescription.

- Passive inhalation of marijuana can, in extreme circumstances, cause urine levels above 100 ng/ml (on screening tests) and/or 15 ng/ml (on confirmatory tests).[40,41,42] The intense exposure required to generate such levels makes it extremely unlikely that someone could test positive merely because others nearby smoked marijuana. In addition, trustworthiness and reliability questions are raised for persons exposing themselves to such an environment, particularly just before scheduled work. DOT advises MROs to reject claims of passive inhalation for positive marijuana metabolite results at or above DOT-mandated cutoffs.[9]

- Several case reports involving infants and toddlers suggest that passive inhalation of freebase cocaine can cause urine levels of benzoylecognine of 300 ng/ml or more.[43,44] In one experiment, an adult had urine levels of benzoylecognine between 8 and 14 ng/ml for 24 hours after secondhand exposure to freebase cocaine smoke.[45] Infants have developed serious neurologic symptoms from riding in poorly ventilated automobiles with PCP-smoking adults.[46] Passive inhalation of freebase cocaine and PCP does occur, but it has not been reported that this can cause urine levels in adults at or above the federal cutoff levels. However, it may cause positive results in nonregulated programs that use lower cutoffs.

- Ingestion of food or drinks laced with drugs can cause positive urine results. The donor may claim that the ingestion was unintentional. Legitimacy, not intent, is at issue. Federal regulations do not address unintentional ingestion; however, the employer's policy may. The MRO's response to these explanations should be consistent. One option is that the MRO can, with the donor's permission, identify such results to the employer as "unintentional positives."

The MRO does not determine, based on the drug test alone, if the person was impaired or if he/she is medically fit for duty. On a stand-alone basis, positive workplace drug tests establish recent drug use, but do not readily correlate with current impairment. (By contrast, alcohol impairment is legally defined at specific blood alcohol concentrations.) While workplace drug tests are not used to indicate past impairment, they are used as a predictor of future impairment.

In some situations, the MRO may also be the certifying physician who determines if the examinee is medically fit. This is a separate and distinct task from drug testing. A drug test can be part of a medical fitness evaluation. For example, FHWA medical standards state that a driver who uses certain controlled substances is medically unqualified. FHWA regulations allow but do not require the examining physician to consider drug test results when evaluating a driver's medical qualification.

Step 5. Report Results to the Employer

The MRO or MRO's staff reports the outcome of each test—negative, verified positive, or invalid—to the employer's contact(s) in accordance with the employer's policy. Reports should provide only the information needed for work placement purposes or regulatory compliance.[47] Identification and concentrations of the drugs detected is, therefore, in the author's opinion, usually unnecessary, except where required by law. For example, DOT/FHWA requires identification of the drug(s) for which a test is verified positive.[48] DOT prohibits reporting of quantitative levels to the employer except in a lawsuit, grievance, or other challenge initiated by the donor.[49] NRC prohibits reporting of levels to the employer except in an appeals process.[50]

In DOT/FRA-mandated testing, a donor with a verified positive result must be provided with a copy of the DOT chain of custody form copy 2. This copy must be sent to the donor within 24 hours following any adverse action.[51]

In DOT/FAA-mandated testing, if the MRO confirms that a donor with a verified positive result has, or has applied for, a Part 67 medical certificate, the MRO must also report the result to the Federal Air Surgeon.[52] (Part 67 certificates are required of private pilots, commercial pilots, airline transport pilots, student pilots before their first solo, flight instructors, flight engineers, and non-FAA, nonmilitary air traffic controllers.) If the donor holds a Part 67 medical certificate, the MRO must determine whether that individual is drug-dependent or nondependent. The Federal Aviation Administration defines drug dependence as "addicted to or dependent . . . as evidenced by habitual use or a clear sense of need for the drug." Figure 5-9 presents a format for reporting positive results to the Federal Air Surgeon.

The DOT/Coast Guard regulations require the employer to report any individual who fails a drug test and who holds a license, certificate of registry, or merchant mariner's document to the nearest Coast Guard Officer in Charge, Marine Inspection (OCMI).[53] The MRO may provide this reporting function for the employer.

Multisite employers and consortia may have site-specific and/or employer-specific contacts. As more site-specific or employer-specific contacts are designated, the effort involved in reporting results increases. Less effort is involved when each site or employer has alternate contacts.

Results are reported by telephone, electronically, and/or in writing. Verbal reports can be misunderstood; the recipient should be asked to repeat each report for verification purposes. As an additional safeguard, verbal reports should be followed up with electronic or written reports. The federal regulations, as of 1991, do not require written reporting from the MRO to the employer. But, written reporting requirements will probably be added to drug testing regulations in the future.

The MRO should record his or her determination in the file of each laboratory-positive drug test. On DOT-type forms, MROs typically record the determination in the MRO block of Copy 2 (see Figure 5-10). DOT does not require the MRO to complete this block; those who do not should record their determinations elsewhere in each drug test's file.

The laboratory result is recorded on the laboratory printout and on copy 2 of DOT COC forms. Either of these can be copied and sent to the employer as a report of a verified positive test (unless the report identifies levels of drugs). This is not advisable, though, because it creates the chance of accidentally sending a laboratory printout or copy 2 of a laboratory-positive/MRO-negative result to the employer. It also distinguishes the laboratory-positive/MRO-negative results to the employer, since they would be reported with a form that differs from the laboratory-negative reports. Therefore, the MRO should use a separate letter or form (e.g., as in Figure 5-11) for reporting every result. Transmissions should be secure

TO: Federal Air Surgeon (AAM-220)
Federal Aviation Administration
800 Independence Avenue, S.W.
Washington, D.C. 20591

REPORT OF VERIFIED POSITIVE DRUG TEST RESULT:
FAR PART 67 CERTIFICATE HOLDER

In compliance with the provisions of Section VII.C of Appendix I, FAR Part 121, I am notifying you of a verified positive result on the following individual who holds an FAA medical certificate issued pursuant to FAR Part 67.

On <DATE>, I, as Medical Review Officer (MRO) for <EMPLOYER>, verified that a drug test performed on <DONOR'S NAME> employed by or being considered for employment by <EMPLOYER> at <SITE LOCATION>, was a positive result for <NAME OF SUBSTANCE(S)>. The specimen was collected on <COLLECTION DATE>; the specimen was analyzed by <NAME OF LABORATORY>; and the employer was informed of the verified positive test result on <NOTIFICATION DATE>.

REPORT OF DEPENDENCE OR NONDEPENDENCE

(CHECK ONLY ONE)

☐ *(Nondependence).* After thorough review of this test in accordance with the DOT Procedures [40 CFR §40] and FAA regulations [14 CFR §121, Appendix I], I have made a determination of nondependence and have recommended that (he/she) be returned to duty subject to the successful completion of the appropriate program of rehabilitation. The individual understands that (he/she) must also pass a return-to-duty drug test and undergo unannounced drug testing for at least 12 months. I am enclosing the supporting documentation that I used in reaching this determination.

☐ *(Dependence).* After thorough review of <DONOR'S NAME> pertinent medical records, and interview with (him/her), and a professional assessment by <NAME AND TITLE>, I have made a determination of probable drug dependence. In compliance with the provisions of Appendix I, FAR Part 121, I am enclosing all documentation pertaining to this individual's test result and my decision for your review and determination.

☐ *(No Determination).* I am unable to make a determination of dependence or nondependence.

MRO: _____ (Print) _____ (Signature)
_____ (Date)

FIGURE 5-9. Report of verified drug test result: FAR Part 67 certificate holder. *Source:* Adapted from Federal Aviation Administration.[54]

> I have reviewed the laboratory result for the specimen identified by this form in accordance with applicable Federal requirements. My final determination/verification is:
>
> (Check one) ☐ NEGATIVE ☐ POSITIVE
>
> SIGNATURE OF MEDICAL REVIEW OFFICER: _____ DATE: _____

FIGURE 5-10. MRO block on DOT-type chain of custody forms.

and confidential. Envelopes containing hard-copy reports should be labeled "PERSONAL AND CONFIDENTIAL," for receipt only by the employer contact or that contact's designated support staff. Facsimile and other electronic transmissions should be sent to terminals where access is similarly limited.

Step 6. Store the Results

Just one federal regulation addresses MRO record retention: DOT/FHWA requires that the MRO be sole custodian of individual test results, and maintain these for at least 5 years.[55] Although it is generally not required, the MRO is often custodian of at least part of the employer's drug testing records by default.

Individual drug test records consist of:

1. Laboratory printouts.
2. Chain of custody forms. If DOT-type forms are used, the MRO should retain copy 3 and a certified, original copy 2.
3. For each laboratory-positive result: an account of the review process, including communications with the donor and the donor's medical provider(s), and a summary of the basis for the determination.
4. Documentation of reporting, for example, an annotated copy of the written report indicating the date and to whom it was sent.

Drug test records may be stored with, or apart from, other employer-sponsored medical records. Storing them separately helps preserve the distinction between drug tests and medical examination records. Under federal regulations, records of positive results generally must be maintained for 2-5 years. Records of negative results generally must be maintained for at

REPORT OF DRUG TEST RESULTS

Donor's Name: _____

Employer: _____

Reason for test: ☐ Pre-employment   ☐ Random   ☐ Post-accident
                 ☐ Periodic   ☐ Reasonable Cause
                 ☐ Other (specify) _____

Collection date: _____
                  Month   Day   Year

Collection site: _____

The laboratory
analyzing the specimen: _____

Test results -- check only one   NEGATIVE _____

                                          or

                                 VERIFIED POSITIVE* _____

                                          or

                                      INVALID _____

Medical Review Officer: _____ (print) _____ (signature)

                         _____ (date)

_____
*i.e., the donor was given an opportunity to discuss the result, and no legitimate explanation was found.

FIGURE 5-11. Report of drug test results.

least 1 year. Records of tests undergoing appeal or judicial action must be kept for the duration of that appeal or action.

The donor or donor's designated representative must have access to the donor's drug test records. Procedures for obtaining access to these records are determined by the employer. A request for records of a verified positive drug test may indicate a legal threat, and should therefore be brought to the employer's attention. The employer, not the MRO, is responsible for responding to any requests for records from a court of law.

ALCOHOL TESTING

The MRO's role in alcohol testing is described by DOT's Coast Guard and FRA regulations, and by NRC's regulations. In each of these rules, a 0.04 percent blood alcohol concentration (BAC) is the cutoff for determining whether a worker is "under the influence." The MRO reviews alcohol test results for potential legitimate medical explanations, for example, ingestion of alcohol-containing cough syrup. The MRO's determination and the BAC are both reported to the employer. In certain settings (e.g., post-accident) employers consider BAC levels extrapolated to a time in the past to determine if the donor was then impaired. The laboratory toxicologist or the MRO may be asked to perform these extrapolations.

REHABILITATION AND RETURN-TO-WORK DETERMINATIONS

Although the MRO role as defined by federal guidelines does not include rehabilitation of the drug abuser, the employer may turn to the MRO, as an expert in substance abuse, for assistance in these efforts. The employer's substance abuse policy and procedures determine who is primarily responsible for evaluation, treatment, and referral of drug-using employees; this is usually assigned to an employee assistance program (EAP). The MRO may serve as a medical consultant to the employer's rehabilitation counselor regarding medical information related to substance abuse.

Several DOT agencies—FAA, Coast Guard, and the Research and Special Programs Administration (pipelines)—ask the MRO to recommend when a worker may return to a security- or safety-related position after failing or refusing to take a drug test, if the employer offers such an opportunity. In making the recommendation, DOT advises the MRO to:[9]

1. Obtain from the rehabilitation counselor an assessment that includes the nature and degree of the person's past substance abuse, progress

in any rehabilitation effort, and the prognosis and recommendations concerning recommended aftercare services;
2. Verify that the person complied with the conditions and requirements of the rehabilitation program in which he or she participated;
3. Determine that a return-to-work urine specimen shows no evidence of current drug use.

If reinstatement is recommended, the MRO must then establish a schedule for random, unannounced drug tests based on the assessment and recommendations of the rehabilitation counselor. The MRO determines the frequency of such testing, which may be in effect for up to 60 months.

SUMMARY

The MRO's primary functions are to review and interpret positive test results, and to assure that the chain of custody forms and laboratory data are in order. Great care must be exercised in executing this function because the outcome can have immediate, decisive consequences. The MRO, as gatekeeper of drug test results, is in theory the ultimate fail-safe in workplace drug testing.

Acknowledgments
The author thanks Melissa Allen, Kenneth Chase, M.D., Mark Upfal, M.D., and Robert Willette, Ph.D., for critically reviewing drafts of this chapter.

References
1. Department of Health and Human Services (DHHS). Mandatory guidelines for federal workplace drug testing programs. *Fed. Reg.* 1988;53(April 11): 11,970–11,989.
2. U.S. Department of Transportation. Procedures for transportation workplace drug testing programs. *Fed. Reg.* 1989;54(Dec. 1):49,854–49,884.
3. The role of the medical review officer (MRO). In B. S. Finkle, R. V. Blanke, and J. M. Walsh (eds.), *Technical, Scientific and Procedural Issues of Employee Drug Testing*. Rockville, Md.: National Institute on Drug Abuse; 1990:29–31.
4. Quality Assurance in the Private Sector Drug Testing Act of 1991. S. 2008, 102nd Congress, 1st Sess., 1991.
5. Autry, J. H. *Advisory 4: Permanent Record Book and MRO Review*. Rockville, Md.: National Institute on Drug Abuse; October 25, 1991.
6. Nuclear Regulatory Commission. Fitness-for-duty programs. *Fed. Reg.* 1989;54(June 7):24,468–24,508.
7. Specimen collection and reporting results. In B. S. Finkle, R. V. Blanke, and

J. M. Walsh (eds.), *Technical, Scientific and Procedural Issues of Employee Drug Testing.* Rockville, Md.: National Institute on Drug Abuse; 1990:21-24.
8. National Laboratory Certification Program. *Laboratory Inspection Report.* Research Triangle Park, N.C.: Research Triangle Institute; 1991:3-K.
9. U.S. Department of Transportation. *Medical Review Officer Guide.* Washington, D.C. 1990.
10. Knisely, R. *Operating Guidance for DOT Mandated Drug Testing Programs.* Washington, D.C.: U.S. Department of Transportation; May 29, 1991.
11. Klonoff, D. C., and A. H. Jurow. Acute water intoxication as a complication of urine drug testing in the workplace. *JAMA* 1991;265:84-85.
12. 49 CFR § 391.97(c).
13. 49 CFR § 391.41(b)(1).
14. 14 CFR § 67, Subpart A.
15. American College of Occupational Medicine. Code of Ethical Conduct for Physicians Providing Occupational Medical Services. Arlington Heights, Ill.: July 23, 1976.
16. Hoyt, D. W., R. E. Finnigan, T. Nee, T. F. Shults, and T. J. Butler. Drug testing in the workplace: Are methods legally defensible? *JAMA* 1987;258:504-509.
17. Hawks, R. L. Analytic methodology. In R. L. Hawks, and C. N. Chiang (eds.), *Urine Testing for Drugs of Abuse.* Rockville, Md.: National Institute on Drug Abuse; 1986:30-42.
18. Bralliar, B. B., B. Skarf, and J. B. Owens. Ophthalmic use of cocaine and the urine test for benzoylecgonine. *New Engl. J. Med.* 1989;320:1757.
19. Cruz, O. A., J. R. Patrinely, G. S. Reyna, and J. W. Kline. Urine drug screening for cocaine after lacrimal surgery. *Am. J. Ophthalmol.* 1991;111:703-705.
20. Hawks, R. L., and C. N. Chiang. Examples of specific drug assays. In R. L. Hawks and C. N. Chiang (eds.), *Urine Testing for Drugs of Abuse.* Rockville, Md.: National Institute on Drug Abuse; 1986:85-92.
21. Dackis, C. A., A. L. C. Pottash, and M. S. Gold. Persistence of urinary marijuana levels after supervised abstinence. *Am. J. Psychiatry* 1983;139:1196-1198.
22. Ellis, G. M., M. A. Mann, B. A. Judson, N. T. Schramm, and A. Tashchian. Excretion patterns of cannabinoid metabolites after last use in a group of chronic users. *Clin. Pharmacol. Ther.* 1985;38:572-578.
23. Karoum, F., L. W. Chuang, T. Eisler, D. B. Calne, M. R. Leibowitz, F. M. Quitkin, D. F. Klein, and R. J. Wyatt. Metabolism of (−) deprenyl to amphetamine and methamphetamine may be responsible for deprenyl's therapeutic benefit: A biochemical assessment. *Neurology (NY)* 1982;32:503-509.
24. Siegel, R. K., M. A. El Sohly, T. Plowman, P. M. Rury, and R. T. Jones. Cocaine in herbal tea. *JAMA* 1986;255:40.
25. Schwartz, R. H., M. Altieri, and S. Bogema. Topical anesthesia using TAC (tetracaine, adrenalin, and cocaine) produces "dirty urine." *Otolaryngol. Head Neck Surg.* 1990;102:200-201.
26. 21 CFR § 1306.32.
27. El Sohly, M. A., and A. B. Jones. Morphine and codeine in biological fluids: approaches to source differentiation. *Forensic Sci. Rev.* 1989;1:14-21.

28. Cone, E. J., P. Welch, B. D. Paul, and J. M. Mitchell. Forensic drug testing for opiates, III. Urinary excretion rates of morphine and codeine following codeine administration. *J. Anal. Toxicol.* 1991;15:161-166.
29. Elliott, H. W., K. D. Parker, M. Crim, J. A. Wright, and N. Nomof. Actions and metabolism of heroin administered by continuous intravenous infusion to man. *Clin. Pharmacol. Ther.* 1971;12:806-814.
30. 10 CFR § 26, Appendix A, Subpart B, 2.7(f)(5).
31. Autry, J. H. *Technical Advisory to All HHS/NIDA Certified Laboratories.* Rockville, Md.: National Institute on Drug Abuse; March 11, 1991.
32. 14 CFR § 121, Appendix I, VI.C.
33. 49 CFR § 199.17(b).
34. 49 CFR § 219.709(b).
35. 10 CFR § 26.24(e)
36. 49 CFR § 391.97(a).
37. 49 CFR § 219.103(a).
38. 10 CFR § 26,Appendix A, 2.9(e).
39. 49 CFR § 219.707(b).
40. Perez-Reyes, M., S. Diguiseppi, A. P. Mason, and K. H. Davis. Passive inhalation of marijuana smoke and urinary excretion of cannabinoids. *Clin. Pharmacol. Ther.* 1983;34:36-41.
41. Cone, E. J., R. E. Johnson, M. Darwin, B. P. Paul, and J. Mitchell. Passive inhalation of marijuana smoke: Urinalysis and room air levels of delta THC. *J. Anal. Toxicol.* 1987;11:89-96.
42. Law, B., P. A. Mason, A. C. Moffat, L. J. King, and V. Marks. Passive inhalation of cannabis smoke and urinary excretion of cannabinoids. *Clin. Pharmacol. Ther.* 1983;34:36-41.
43. Heidemann, S. M., and M. G. Goetting. Passive inhalation of cocaine by infants. *Henry Ford Hosp. Med. J.* 1990;38:252-254.
44. Bateman, D. A., and M. C. Heagarty. Passive freebase cocaine ("crack") inhalation by infants and toddlers. *Am. J. Dis. Child.* 1989;143:25-27.
45. Baselt, R. C., D. M. Yoshikawa, and J. Y. Chang. Passive inhalation of cocaine. *Clin. Chem.* 1991; 37:2,160-2,161.
46. Schwartz, R. H., and A. Enihorn. PCP intoxication in seven young children. *Pediatr. Emerg. Care* 1986;2:238-241.
47. American College of Occupational Medicine. Drug screening in the workplace: Ethical guidelines. *J. Occup. Med.* 1991;33:651-652.
48. 49 CFR § 391.87(e).
49. 40 CFR § 40.29(g)(3)
50. 10 CFR § 26, Appendix A, Subpart B, 2.7(g)(3).
51. 49 CFR § 219.707(c).
52. 14 CFR § 121, Appendix I (VII)(C)(5)(c).
53. 46 CFR § 16.201(c).
54. Federal Aviation Administration. *Aviation Medical Review Officer Guide.* Washington, D.C. 1990.
55. 49 CFR § 391.87(d).

6
Risk Management

William J. Judge, J.D., LL.M., and
Robert Swotinsky, M.D., M.P.H.

A positive drug test can immediately affect a person's career and reputation. Because of these repercussions, that person may take legal recourse against any party involved in this testing, even if the test was done correctly. A test that fails to detect drugs or alcohol use can also have legal repercussions if the employee subsequently has an accident attributed to drugs or alcohol. Liability is a major concern of employers and of providers of drug testing services.

Drug testing is an evolving and relatively new area, with uncertain risks to the employer and to the provider of drug testing services. Lawsuits against providers of drug testing services have, to date, been infrequent. This is in part because, historically, most testing was done on applicants who, if they tested positive, were not hired. Starting in the 1990s, more testing has been done on employees who, if they test positive, may be fired. Being fired typically causes greater resentment and financial loss than not being hired. The prospect of recovering higher damages is an incentive that makes the fired worker, rather than the applicant who is not hired, a more likely plaintiff. The increase in in-service testing may pose a greater risk of liability to providers of drug testing services.

This chapter suggests strategies by which providers can lower their risk of liability. These strategies primarily address Medical Review Officers (MROs), but also address concerns of collection sites, laboratories, and employers. Although there can be no guarantee that lawsuits will not occur, these strategies are designed to help decrease the risk of a lawsuit and to increase the likelihood of success if one is sued.*

---

*This chapter is not intended as a substitute for legal advice, but instead as a general discussion of the subject matter. Neither the publisher nor the editor is in the business of providing legal advice, for which an attorney should be consulted.

An understanding of the legal framework in which drug testing is performed is also essential. The MRO, in particular, must recognize the legal limits of drug testing. The role of the MRO in drug testing programs provides a "check" on that process, but also places the MRO at the focus of potential liability.

THE LEGAL FRAMEWORK OF DRUG TESTING: THE RISK

Drug testing is a sensitive issue because those who test positive may lose their jobs. But, because drug or alcohol use can pose significant problems at work, testing may be appropriate. While the association between substance use and on-the-job injuries is largely anecdotal, it would be difficult to deny that substance abuse can impair one's ability to safely perform assigned tasks. One example is tragically illustrative: In 1987 outside Chase, Maryland, a Conrail/Amtrak train accident killed 16 people and injured scores of others. The Conrail engineer admitted smoking marijuana just before the collision.[1]

In part as a response to this disaster, and as a result of a growing recognition of the negative impact of substance abuse on workplace safety, federal legislation has been issued that requires certain employers to institute drug testing programs. At the same time, some states and municipalities have enacted laws that limit workplace drug testing. These laws and regulations, combined with existing laws that protect the rights of employees and injured third persons, have created a complex, sometimes conflicting, legal framework.

Compliance with these laws, where applicable, offers some protection against a claim that testing was undertaken negligently or recklessly. The federal drug testing regulations provide highly detailed drug testing procedures and standards that help assure accuracy, uniformity, and confidentiality. Following these procedures, particularly where required by law, can help mitigate the risk in performing drug testing.

The risks in performing drug testing are being defined through case law from suits filed by employers, unions, and injured third parties against providers. Liability may accrue to the provider in the following circumstances:

1. Negligence in collecting or preserving specimens, conducting analyses, or reviewing results;
2. Improper disclosure of positive test results;
3. Failure by the MRO to inform the employer of potential safety risks, other than the drug test result, learned while reviewing a laboratory-positive drug test.

The provider may be a co-defendant in allegations against the employer, for example:

1. Claims of defamation of character, invasion of privacy, and various public policy violations;
2. Failure to take appropriate action to remove an employee from a safety-sensitive position when that person had a verified positive drug test result;
3. Failure to comply with collective bargaining agreements or other contractual obligations when taking adverse employment action based on a drug test result.

If a false-positive test outcome is alleged, the donor's suit is likely to include claims for defamation, claims for infliction of emotional distress, and claims seeking both lost wages and punitive damages. The employer may in turn bring an action (breach of contract) against the MRO to shift the burden of liability or to recover damages paid as a result of the MRO's negligence.

To successfully maintain legal action against a provider of drug testing services, the donor, employer, or injured third party must prove that the provider owed him or her a duty, that the duty was breached, that injury occurred, and that the provider's breach of duty caused that injury. How is that duty created? It may be the product of an express or implied contract, may be statutorily created, or may be the product of judicial interpretation known as "common law."

To limit personal liability or to protect other business interests, a physician serving as an MRO may wish to incorporate the MRO service in accordance with the laws of the state in which he or she does business. In addition to the usual limitation on liability,* there may be advantages in obtaining liability insurance coverage.

ASSISTING CLIENTS IN PROGRAM DEVELOPMENT

The MRO function is but one link in a chain of activities in the drug testing process. This chain starts with the employer and includes the donor, the collection site, the laboratory, the MRO (in some programs), a third-party program administrator (in some programs), and then the employer again.

---

*Check state laws carefully. In some states professionals such as lawyers and physicians cannot shield themselves from personal professional liability through the use of corporations.

Y/N	Does the client have a drug/alcohol testing policy?
Y/N	Do procedures exist that will assist the employer in carrying out the day-to-day operations of the program?
Y/N	Do the procedures include handling collection problems?
Y/N	Must the client follow federal or state laws/regulations?
Y/N	Is a union involved? If so, have negotiations taken place? (It is essential.)
Y/N	Do the procedures indicate how problems that involve the MRO function (e.g., contacting hard-to-reach donors) will be resolved?
Y/N	Is a confidential reporting system in place?
Y/N	Do all involved recognize the need for confidentiality?

FIGURE 6-1. Critical questions for evaluating an employer's drug testing program.

Deficiencies at any level can put all involved at greater risk. Such risks are avoidable.

Providers of drug testing services should review the employer's policy and procedures for technical adequacy, regulatory compliance (where applicable), and ethical* and employee relations considerations. Figure 6-1 presents questions that should be answered during such a review. It is likely that the MRO's advice will be sought in establishing and maintaining aspects of the drug testing program. The MRO should, however, recognize that such advice may fall outside the scope of work for which he or she has been retained. The employer-MRO contract should clearly define the scope of work provided by the MRO. Acting beyond that work statement increases one's risk.

RISK MANAGEMENT STRATEGIES

Use of Contracts

Use of written contracts to define engagements with clients may be the most critical strategy in managing the MRO's risk. The contract can help in sev-

---

*Ethical guidelines developed by the American College of Occupational Medicine[2] (see Appendix A) present features that should be included in any workplace drug testing program.

eral ways. First, it defines the role the MRO plays in the particular employer's program. Second, the MRO and the employer are compelled to address and resolve certain issues in developing a contract. Third, by clearly defining the scope of the engagement and the responsibilities to be assumed, a contract may help the MRO obtain credit or insurance. Finally, the contract provides a statement of work that, if written properly and adhered to, limits the MRO's risk. If the statement of work does not call for service in a particular subject area, the employer is less likely to ask for this service, and the MRO is less likely to provide it. The contract's statement of work can serve as a defense in the event of a suit that alleges injury due to the MRO's inaction in an area that falls outside of the contract. Figure 6–2

- ☐ Identify the Parties.
- ☐ Define the Term of Contract; for example, 6 months, 1 year, etc.
- ☐ Define the MRO's Duties:
  - Be specific.
  - Role limited to review of test results?
  - Role that includes assistance with rehabilitation?
  - Role that includes assistance with rehabilitation and determinations of fitness for return to duty.
- ☐ Both parties agree to abide by local, state, federal laws and regulations.
- ☐ Schedule of fees. (Be specific!)
- ☐ Insurance Issues:
  - Does your existing policy cover the MRO function? Is a rider to that policy required?
  - Can you be covered as an additional insured party under the Client's insurance policy?
  - Does your fee cover the added cost of insurance?
- ☐ Client will make necessary personnel available.
- ☐ Procedures to maintain confidentiality.
- ☐ Procedures for determining invalid tests:
  - What constitutes an "invalid test"?
  - Will the laboratory and/or MRO be responsible for reviewing the adequacy of laboratory-negative results?
- ☐ Who pays for reanalyses?
- ☐ Which state laws apply?

FIGURE 6–2. The MRO new client setup checklist. This is a guide only. Seek the advice of the appropriate professionals before entering into any agreement.

presents a checklist of steps the MRO should consider in establishing a new client relationship.

The contract should make clear that the provider may furnish and interpret test results, but only the employer decides whether to hire or fire the worker. A typical clause might provide:

> The parties understand and agree that (PROVIDER) will not make any decisions regarding the client's employment, termination, retention, or disciplining of any employee, former employee, or applicant for employment, and that the client shall have sole responsibility for all such decisions.

Similarly, because the decision to conduct drug testing is made by the client, not by the provider, the provider should not accept exposure to liability based on the alleged impropriety of that decision. The contract should require the client to indemnify the provider against claims that the drug testing program is in any way invasive of privacy or otherwise contrary to law or public policy. The provider cannot, however, obtain indemnity for his/her own negligence in performing the contract.

Thorough Documentation

Drug testing is paper-intensive. Providers of drug testing services should be meticulous in their record keeping. Each test review should be documented in detail. This documentation should permit a plaintiff's attorney to readily understand the process and reasoning employed in reaching your decision. Each record is a potential courtroom exhibit. Furthermore, good records may persuade a plaintiff's attorney that no case exists.

Standardization

Adherence to standard operating procedures helps assure consistent treatment of each donor, and makes any potential courtroom testimony based on these procedures more credible. These procedures should not distinguish between applicants and current employees. Each donor's test should be handled with an equally high degree of uniformity, reliability, and confidentiality.

Confidentiality

The most likely subject of a lawsuit, and yet the most avoidable, is confidentiality. A breach of confidentiality may give rise to a claim based on a breach of contract or "invasion of privacy." All information obtained in the course of testing is sensitive and must be protected as confidential medical information. As a way of reducing the risk, a donor should be asked to sign a release prior to each collection. The release should allow all those in the chain of drug testing—collector, laboratory personnel, MRO, and employer—access to all relevant information. This information must not be shared with anyone else without the prior written consent of the donor unless required or permitted by law. Sharing such results with law enforcement authorities would in all likelihood be futile. Except in the state of Massachusetts, the *use* of drugs is not a crime; it is the *possession* of a controlled substance that is criminal. A positive drug test result shows past use, not possession, of the drug. Therefore, the test result is not evidence of a crime.

Open Communication

A donor with a verified positive test may contact the MRO or the employer to discuss the test after its outcome appears final. If the donor is turned away or remains unsatisfied with the result, he or she may seek satisfaction by means of a lawsuit. To limit the possibility of this scenario, the MRO would be wise to allow the donor as much latitude in explaining the result as possible without breaching contractual, regulatory, or professional obligations. Be open to the donor's request to provide documentation that might explain the result; pursue all reasonable explanations.

The MRO should keep in mind the scope of his/her engagement for the employer. No physician–patient relationship is ordinarily created. But if the MRO expands his/her role beyond the interests of the employer and begins to advise or "treat" the donor, it may be argued that a physician–patient relationship has been created, thereby entailing the potential for medical malpractice allegations.

Maintaining a Professional Demeanor

Donors should be treated with respect by all staff involved in the drug testing process. Some substance abusers, especially under these sensitive circumstances, can be difficult. The risk of adverse reaction is reduced by responding with a professional demeanor. Moreover, a professional approach helps sustain dialogue with the donor and reinforces the credibility of the procedures.

Continuing Education

Workplace drug testing is a relatively new and rapidly evolving phenomenon. The regulations and the scientific literature regarding drug testing have not been effectively disseminated to employers and providers. To fill this void, federal agencies such as the Federal Aviation Administration and the Federal Highway Administration, and certain private groups such as the

American College of Occupational and Environmental Medicine and the American Society of Addiction Medicine, have offered seminars to employers and to providers of drug testing services. As yet, no standard curriculum for MROs exists. There is a recognized need for continuing education of MROs.[3] Education is key to managing the risks of drug testing.

Obtaining Adequate Insurance Coverage

Regardless of the steps taken to reduce the risks of service as an MRO, lawsuits will occur and losses will be incurred. Adequate insurance coverage must be obtained. The key is finding the right coverage for the function to be performed. The MRO should review his or her coverage to be certain that, in the event of a claim, the policy covers the MRO function. Consultation with appropriate professionals who are willing to learn what the MRO function entails is strongly recommended. One should not wait until a claim is made to explore the adequacy of insurance coverage.

SUMMARY

Drug testing is risky business. Providers of drug testing services find themselves in a new and evolving field. In many respects, the level and extent of liability is yet unknown. However, the risks are not only manageable but are, for the most part, avoidable. Federal employee drug testing regulations contain excellent procedures that providers should, for the most part, follow, even where not legally applicable.

Qualified, professional individuals are needed if any drug-free initiative is to succeed. Recognizing the risks and taking appropriate steps to reduce those risks will allow all involved in the process to accomplish the goal of a healthier work force in a healthy workplace.

Acknowledgments
The authors thank Ronald Precup, Esq., and David Copus, Esq., for critically reviewing drafts of this chapter.

References
1. Gust, S. W., and J. M. Walsh. Research on the prevalence, impact, and treatment of drug abuse in the workplace. In S. W. Gust, and J. M. Walsh (eds.), *Drugs in the Workplace: Research and Evaluation Data*. Rockville, Md.: National Institute on Drug Abuse; 1989:3–13.
2. American College of Occupational Medicine. Drug screening in the workplace: Ethical guidelines. *J. Occup. Med.* 1991;33:651–652.
3. The role of the medical review officer (MRO). In B. S. Finkle, R. V. Blanke, and J. M. Walsh (eds.), *Technical, Scientific, and Procedural Issues of Employee Drug Testing*. Rockville, Md.: National Institute on Drug Abuse; 1990:29–31.

7
Employee Assistance Programs

Larry V. Stockman, Ph.D.

This chapter serves to help medical departments, Medical Review Officers (MROs), human resource and benefits managers, and consultants understand, work with, and evaluate employee assistance programs (EAPs). Federal regulations feature the EAP as a required component of a comprehensive drug-free workplace program.[1,2] An EAP serves to assist employees with drug abuse or other problems by means of counseling, treatment, or referral to more specific centers. EAPs are a forum through which employers can offer education and treatment to workers with drug dependency problems.

BRIEF HISTORY

Employee assistance programs have a recent but complex history in the United States.* Following the establishment of Alcoholics Anonymous in 1935, a series of companies began occupational alcoholism programs (OAPs): New England Electric, Kaiser Shipbuilding, Du Pont, and New England Telephone.

World War II brought on a new demand for OAPs. Both unions and employers were under increased pressure to meet the production demands of the war effort. They recognized that substance abuse affected worker productivity. As a result, more employers began OAPs, which continued to expand throughout the forties and fifties.

---

*Macy's Department Store began an employee assistance program in 1917. Its purpose was to help employees with substance abuse problems.

In the 1960s, OAPs were expanded within unions, companies, and federal/state agencies to help troubled employees with emotional, family, and mental health problems. As drug abuse became a serious problem in the 1970s, and as psychological problems (i.e., domestic violence, divorce, and so on) were seen to negatively affect work performance, OAPs were expanded into more comprehensive programs known as "broadbrush" or comprehensive mental health-care programs, and later as EAPs.

Since the early 1970s, the number of companies with EAPs has increased dramatically, and the scope of assistance provided by EAPs has expanded significantly. Less than two-thirds of Fortune 500 companies had implemented EAPs by the beginning of the eighties. It was during the eighties, however, that companies began recognizing the heavy impact of personal problems on performance and productivity. Corporate leaders began to speak out about EAPs "making good business sense."[3]

The rapid growth of specially designed mental health/substance abuse cost-control programs, such as managed mental health-care (MMH) providers, using the EAP as a control (gatekeeper) for admissions into mental health-care clinics and hospitals that offer reduced fees, has brought new emphasis to EAPs. Managed health-care plans proliferated in the late 1980s. Many of them have offered an EAP component as part of their array of services.

EAPs, other mental health professionals, and employers have collaborated to develop EAP/MMH programs that provide quality attention and emotional health to all employees and their family members. The goal is to provide a program that acts as an advocate for the employee/dependent in order to maximize the therapeutic action of mental health benefits.

The impact of cocaine, marijuana, and alcohol on workplace productivity, medical insurance costs, and safety became widely recognized in the 1980s. In response, the federal government passed multiple regulations in the late 1980s. The Drug-Free Workplace Act of 1988 requires that certain federal contractors and grantees inform employees about "any available counseling, rehabilitation and employee assistance programs."[4] Additional federal regulations in the late 1980s introduced the MRO (physician) role in workplace drug testing, a role that can encompass particular aspects of employee assistance. EAPs were adopted by an increasing proportion of employers. By 1990, 87.5 percent of Fortune 500 companies had implemented, or were in the process of implementing, an EAP (compared to 25% in 1972 and 57.7% in 1979).[5] Small- and medium-sized employers have been slower to adopt EAPs. A 1991 survey found only 48 percent of employers with an EAP in place, and 10 percent of employers intending to begin offering EAP services within the year.[6]

DESCRIPTION

Employers provide EAPs to employees and their immediate family members at no direct cost to the employees. The employer pays for the program, either by direct payroll for an internal program or by contract for externally managed programs. Since no payments need be made by the employee, confidentiality is assured and the program remains voluntary. Special toll-free and local telephone numbers are made available. This emphasizes the desire to maintain confidentiality. A specific number of counseling sessions is made available for each calendar year.*

Typically, EAPs provide counseling assistance for multiple personal or work problems. Family, marital, and emotional issues are the most frequently addressed issues (see Figure 7-1). Chemical dependency and the resulting co-dependency issues are significant family and work issues that are addressed often within the EAP counseling sessions. The typical distribution of major issues and problems that EAP counselors address is as follows:†

Emotional/psychiatric	28%
Marital	25%
Family	13%
Chemical dependency (employee)	9%
Chemical dependency (family)	8%
Domestic violence	5%
Career/work environment	5%
Medical	2%
Legal	2%
Financial	2%
Other	1%

There are approximately 10,000 EAP providers in the United States.‡ EAPs can be classified as internal EAPs, which are employed by and serve

---

*There are many varieties of EAPs. The most common models today are assessment/referral and short-term therapy models. The traditional program is assessment/referral, which consists of no more than three sessions per employee per year. A more common model, expanded during the 1980s, is an eight-session assessment and brief therapy model.

†These data are based on the author's years of experience as an employee of Human Affairs International, working with employees and their families, and preparing statistical reports for employers.

‡This estimate is based on the author's random selection telephone survey of large-, medium-, and small-size companies. The smaller and more local the company, the more likely that a local counselor provides EAP services. The results of this survey indicate that there are hundreds of local EAPs providing such services in small, as well as large, cities across the nation.

PERSONAL PROBLEMS	WARNING SIGNS (CHANGES IN)
FAMILY MARITAL MEDICAL DRUGS & ALCOHOL FINANCIAL EMOTIONAL	APPEARANCE BEHAVIOR PERSONALITY GRIEVANCES # OF DISCIPLINARY ACTIONS SAFETY VIOLATIONS ACCIDENTS ON/OFF JOB COVER UP

DOCUMENTABLE PERFORMANCE PROBLEMS

LOWERED PRODUCTION
DECLINE IN QUALITY
DISCIPLINARY ACTIONS
MISSED DEADLINES
ATTENDANCE
EXCESSIVE EARLY OUTS
EXCESSIVE TARDINESS
SAFETY VIOLATIONS
INCREASE IN CUSTOMER COMPLAINTS
ACCIDENTS
EXCESSIVE ABSENTEEISM

FIGURE 7-1. Problems addressed by employee assistance programs.

a single employer, or external EAPs, which provide services to multiple employers. A survey of randomly selected employers found that 26 percent used internal EAPs, 71 percent used external EAPs, and 3 percent used both internal and external EAPs.[7]

The OAPs were traditionally internal (in-house) programs. Over the years, for reasons of confidentiality, liability, and cost, many employers have changed from internal programs to outside providers of these services. Following the advent of managed mental health care, an increasing number of employers have designed programs that combine internal and external programs. Most EAP programs are organized along similar lines (see Figures 7-2, 7-3, and 7-4).

EAP EDUCATION AND TRAINING

The Drug-Free Workplace Act of 1988 requires that covered employers provide employees with an educational program regarding the dangers of drug abuse in the workplace and the availability of drug counseling, rehabilitation, and employee assistance programs.[4] A program of this nature is appropriate in any setting. Employees should be instructed on how to access the EAP. Supervisors should be instructed on how to effectively work with the EAP.

The EAP can be a significant and powerful resource to assist in maintaining a drug-free workplace and healthy, productive work force. Union and corporate management can be encouraged to use the EAP through ongoing training and educational programs related to addiction, drugs and the family, stress, and twelve-step programs. An EAP and drug-free workplace program may exist on paper only, a token gesture by the employer, unless supervisors and employees receive ongoing education and training related to substance abuse and other mental health-care problems.

EAP SERVICES RELATED TO DRUG ABUSE

EAPs should be separated from the process of specimen collection and analysis. The federal government's model plan for a drug-free workplace

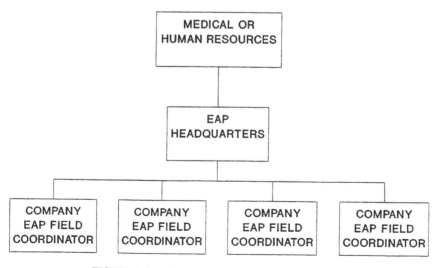

FIGURE 7-2. Internal employee assistance program.

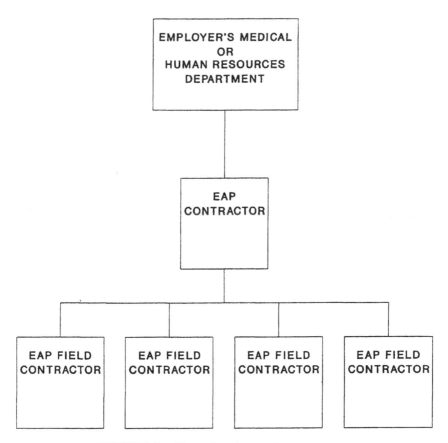

FIGURE 7-3. External employee assistance program.

program specifically prohibits the EAP from involvement in the collection or initial reporting of drug tests.[2] Drug testing is often viewed as punitive; the EAP's mission is to be resourceful, helpful, and curative. The EAP can provide valuable support to the employer and to union representatives, especially where work-related problems negatively affect an employee's work performance or threaten his or her continued employment.

Any employee who tests positive, whether mandated by federal legislation or not, should be offered assistance to obtain counseling support. Employees typically gain access to EAP services through two routes: voluntary self-referrals and (mandatory or "strongly suggested") supervisory referrals. EAP services typically follow these steps:

1. A 24-hour crisis hotline/answering service should be available to employees.

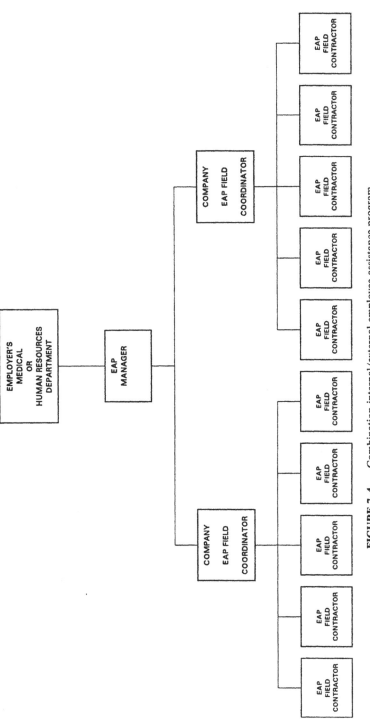

FIGURE 7-4. Combination internal/external employee assistance program.

2. The EAP counselor informs the employer's representative or supervisor of the available appointment time(s). Appointments should be readily available, preferably within 24 hours.
3. The employee is responsible for scheduling the appointment with the EAP.
4. The EAP counselor obtains a written release from the employee, usually during the first visit, that allows the EAP to inform the employer's representative and/or supervisor that the employee has started the therapy process. No other information should be given.
5. The EAP counselor initiates the appropriate treatment: primary/acute care (inpatient, if required), outpatient, Alcoholics Anonymous, Narcotics Anonymous, Cocaine Anonymous, and so on.
6. The EAP counselor monitors the employee's recovery up to and including the employee's return to work.
7. The EAP counselor coordinates a return-to-work (RTW) conference that includes the treatment team, the employer's representative, the counselor, and the employee's immediate supervisor. (The "Monitoring Recovery" section of this chapter describes the RTW conference in detail.)
8. The EAP counselor and employer's representative/supervisors are encouraged to continue communication during the first year of the employee's recovery.
9. Ongoing, random testing of the employee is an important tool to assist the employee in maintaining sobriety and continuing in a successful recovery program.

GUIDELINES FOR REFERRING EMPLOYEES

Most referrals to the EAP are self-referrals. Other sources of referrals include supervisors, physicians, and family members. Drug testing serves primarily as a deterrent to drug abuse. Employees who test positive should be referred to the EAP immediately. The EAP should be told that the referral was prompted by a positive drug test result.

EAPs are designed to provide professional counseling to employees and their family members at no direct cost. Their confidential nature is key to their effectiveness and utilization. The EAP counselors, working with the union and company managements, help assure confidentiality. Even in instances of positive drug tests, confidentiality is an important goal.

EAPs typically receive the following types of referrals:

- *Self-referral.* Of all referrals to EAPs, 80–85 percent are self-referrals. EAPs are designed to function in this way to provide maximum confidentiality. The employee or the employee's family member can use a toll-free telephone number to schedule a counseling appointment directly with the EAP organization. Employers, unions, and EAPs should provide employees and their family members with wallet cards, posters, trifolds, and other information that reminds employees that the program is confidential and free.

- *Supervisory referral.* A supervisor who has an employee with a work performance problem can use the EAP to help the employee remedy the problem (see Figure 7-5). Supervisors may hesitate to refer workers to the EAP for the following reasons: Supervisors may not recognize, and may not have been instructed, that the EAP can help them manage their employees. Supervisors may believe that they, not the EAP, are responsible for handling employee problems. Supervisors may themselves have unresolved personal problems that make them feel compromised in referring employees to the EAP. The supervisor initiates a referral by confronting the employee with the problem, for example:

"Joe/Mary, you have worked with us quite a while now. Your performance up to the past several months has been good. However, something has happened recently. I don't know what it is. It's actually none of my business. However, I am concerned about your work performance. I am responsible for keeping our performance and productivity at a high level.

"Your performance is affecting our performance. I don't want to lose you; the organization doesn't want to lose you. You are a valued employee. I'm concerned about you as a person. What is crucial to me is that you turn around your performance.

"As you may or may not know, we have an EAP. This means that you can go to a quality counselor, with whom you can sit down and discuss whatever personal issues are going on in your life that might be affecting your performance. Here's a wallet card with the telephone number of our EAP. I would like for you to call them and go in to see them within two days. In addition, I would like to meet with you next week to discuss our conversation today and your performance during the forthcoming week. I plan to meet with you consistently over the next six months."

- *Medical referral.* In organizations that have a medical department, 5–10 percent of the referrals to the EAP come from that department.[8] Frequently, the EAP reports functionally to the medical department, which offers well-established procedures to maintain confidentiality and is typi-

150 Employee Assistance Programs

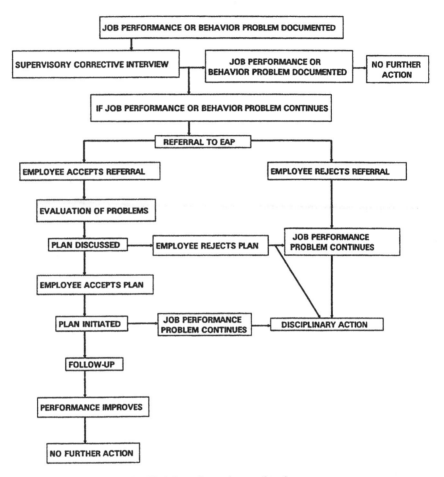

FIGURE 7-5. Supervisory referral process.

cally well respected by employees. The EAP functioning as part of the corporate medical department benefits from these advantages, and is more highly utilized by employees.

• *Mandatory referral.* Where workplace drug testing is conducted, the employer should encourage any employee who tests positive to make an appointment immediately—the same day, if possible—with an EAP counselor. If an employee chooses not to seek help after a positive drug test, the employer must then decide on appropriate measures up to and including termination of employment.

MONITORING RECOVERY

The employee and EAP counselor maintain ongoing communication during recovery, for example, weekly/monthly telephone calls and/or individual counseling sessions. The EAP counselor monitors the recovering employee's attendance at self-help groups and the employee's communication with his or her primary sponsor, and monitors for signs of relapse. The EAP usually continues monitoring the recovering employee for at least a year.

Neither the EAP counselor or the EAP organization should ever be compromised or pressured into an enforcement role. To do so would damage the EAP's credibility to employees, and irreparably destroy trust in the confidentiality of the program.

If the employee enters and actively participates in an effective recovery program, and a RTW conference is held before the employee's return to the workplace, the decision about returning to work should be relatively easy. The RTW conference is the primary tool for assessing the recovering employee's ability to return to the workplace. Figure 7-6 describes, and Figure 7-7 gives guidelines for, the RTW conference.

Participants in the RTW conference typically include the recovering employee and the treatment team (program director, primary therapist, continuing care counselor, physician), the EAP counselor, and the employee's supervisor or medical representative. The RTW conference should occur once the treatment team has decided that the impaired employee is sufficiently advanced in the recovery program.

Figure 7-8 presents the responsibilities of each attendee at the RTW conference. A recovery plan agreement (see Figure 7-9) and a recovery contract (see Figure 7-10, on pages 156-157) should be discussed and signed by the attendees. These provide additional leverage to encourage the employee's recovery. Ongoing, random drug/alcohol testing of the employee, as described in Figure 7-11 (page 158), may also help encourage sobriety.

COST-BENEFIT ANALYSES

Workers with unresolved personal problems cost U.S companies approximately $1 billion each year by absenteeism, on- and off-the-job accidents, errors, sick leave, and health insurance.[9] Research on the cost-effectiveness of EAPs, particularly "broadbrush" EAPs, is scanty. Recognizing this, McDonnell Douglas Corporation in 1985 revised its 14-year-old EAP, after a cost-benefit study revealed that the EAP program was ineffective in managing the substance abuse and mental health problems of its employees. Rather than simply cutting benefits (such short-sighted efforts can cost a company millions of dollars, in the long run), McDonnell Douglas and its EAP management chose to do the following:

> The return-to-work conference:
> - Is a meeting of the employee, the supervisor, the EAP counselor, and the treatment provider. It is usually held at the treatment facility toward the end of the inpatient/outpatient program.
> - Is a forum for the employee to see how the above parties will interface on his/her behalf during the next 12 or more months.
> - Sets up an alliance among the parties regarding expectations of all parties during recovery.
> - Is a mechanism to support the employee's efforts at maintaining recovery.
> - Is a forum for the supervisor to state clearly his/her expectations of the employee when he/she returns to work.
> - Provides the supervisor an opportunity to meet with the EAP counselor and the treatment team and to hear the need of the employee from his or her perspectives.
> - Provides the supervisor an opportunity to share with the EAP counselor information about the employee's performance, in order to aid in continuing assessment and treatment during recovery.
> - Is the occasion during which the recovery plan agreement is discussed and signed by a company representative and the client.
> - Provides the return-to-work recovery plan contract, which gives assurance that the supervisor will receive regular updates on the progress of the employee participating in the recovery program.

FIGURE 7-6. Return-to-work conference definition.

- Assess individual cases carefully, emphasizing quality care;
- Provide the best possible front-end service (although initially more costly);
- Select providers (clinics, hospitals, programs, clinicians) with a track record of cost-effective care;
- Monitor each case, both during treatment and for up to 2 years after the employee returns to the workplace;
- Include the entire family in treatment, as needed.

The McDonnell Douglas results in cost savings are impressive:

- Over a 4-year period, McDonnell Douglas employees who used the EAP for chemical dependency missed 44 percent fewer work days and filed $7,300 less in insurance claims, than those who did not use the EAP.
- Employees treated for chemical dependency and psychiatric health issues, and who did not use the EAP, used 50 percent more in health-care costs than did those who used the EAP.

> - All chemical dependency cases that have resulted from a supervisory or medical referral should make use of a return-to-work (RTW) conference.
> - When an employee is a "voluntary" chemical dependency admission, the primary treatment facility team should offer a RTW conference for the employee. The employee and EAP counselor should discuss the implications of holding the conference.
> - It is expected that the person undergoing treatment return to work immediately after discharge from treatment. There are no days off. A prompt return to the workplace helps recovery.
> - All agreements and contracts should be discussed at length during the RTW conference. These should be signed in the presence of the RTW conference team.
> - The treatment facility is reponsible for informing the EAP counselor of the employee's status during the period of recovery. Regular communication between the facility and EAP counselor should continue for at least one year.

FIGURE 7-7. Return-to-work conference guidelines.

- A total of 40 percent of employees who were treated for drug/alcohol abuse but did not use the EAP left the company within 4 years, as opposed to only 7.5 percent of those who used the EAP.

McDonnell Douglas predicted an estimated $5 million would be saved (over and above the cost of managing the EAP) during the 1990–1992 three-year span, due almost entirely to making effective use of a quality EAP.[10]

In addition to the financial savings, other, less tangible benefits can arise from the EAP, for example, high employee morale, high productivity, improved decision making, high quality of work, and commitment. It is appropriate for employers or unions to schedule cost-benefit evaluations of their EAPs, much like the McDonnell Douglas effort. Comparisons between control groups (employees who have not used the EAP) and groups of employees who have used the EAP can give valuable information. Major variables that merit review are absenteeism, turnover, tardiness, illness (medical issues: stomach, liver, etc.), disability/medical claims, on/off-the-job accidents, and overtime/termination/recruitment expenditures.

(Text continues on page 159.)

SUPERVISOR RESPONSIBILITIES

☐ Receive at least 2 hours of training from the EAP counselor about relating to and managing the employee returning to work.
☐ Assure that all employee–employer agreements and documents are reviewed and signed at the return-to-work (RTW) conference.
☐ Clearly state at the RTW conference the employer's expectations regarding the recovering employee's return to the workplace.
☐ State clearly to the recovering employee the potential consequences of a relapse or the employee's failure to adhere to the prescribed recovery program.

TREATMENT FACILITY RESPONSIBILITIES

☐ Assure the supervisor and recovering employee that an ongoing communication between the facility and the EAP counselor will occur on a regular basis during the first year of recovery.
☐ Present the discharge plan, which defines explicitly the first year's recovery and aftercare program.

RECOVERING EMPLOYEE RESPONSIBILITIES

☐ Understand that sobriety and recovery are forever.
☐ Understand the significance of the RTW conference.
☐ Perceive the value of immediately returning to work after the primary treatment and RTW conference.
☐ Respect and clearly state at the RTW conference the importance of the recovery program and his/her determination to adhere to it.
☐ Clearly state at the RTW conference his/her understanding of the time requirements for adhering to the spirit of recovery.
☐ Review with the RTW conference team his/her understanding of the primary treatment facility's discharge plan, the agreement (contract) with the employer, union, and/or EAP counselor, and how he/she expects to meet these expectations.
☐ Understand the EAP counselor's responsibility to report to the appropriate personnel on the employee's recovery progress.
☐ State as clearly as possible to the RTW conference team his/her dedication to lifelong sobriety within the "day-at-a-time" philosophy.
☐ Review with the RTW conference team the importance of the recovery process.

FIGURE 7-8. Return-to-work conference checklist.

CONFIDENTIAL

The purpose of this agreement is to provide a clear understanding of performance expectations. My recovery plan will extend for ____ months and become effective on _____.

I understand and agree with the recovery contract presented to me on this date. I acknowledge it is my responsibility to follow the counseling and continuing-care plans that are described herein. I also understand I will not receive scheduled time off from my job to satisfy program requirements.

I understand total abstinence from all drgus/alcohol is expected and that any relapse must be communicated to my counselor and my supervisor (or my employer's medical department) immediately. Failure to follow this procedure, failure to follow the recommended counseling and continuing-care meetings of my recovery contract, or performance problems will result in discipline that may include termination of my employment. I will be notified in advance of any changes to the counseling or continuing-care meetings required by my recovery contract.

I have read my employer's Alcohol and Drug Abuse Policy. I understand I will be subject to periodic and unannounced alcohol and drug testing for a period of 60 months. The frequency of testing will be determined on an individual basis taking into consideration my rehabilitation progress and the type of chemical dependency for which I received treatment. I understand that a positive alcohol or drug test result or refusal to submit to periodic testing is grounds for discipline as referenced in the Alcohol and Drug Abuse Policy.

_____   _____
Employee's Signature   Date

_____   _____
Employer Representative's Signature   Date

FIGURE 7-9. Sample recovery plan agreement.

The purpose of this contract is to provide a clear understanding of the recovery expectations, as described below. The contract hereby states:

I am willing to fulfill the following conditions to further my recovery from chemical dependence and reassure my employer of my commitment to recovery:

- I commit to maintain total abstinence from alcohol and all drugs not prescribed by a physician familiar with my state of recovery. I will report any prescriptions to my EAP counselor.
- I commit to maintain myself in treatment for chemical dependency for a minimum of ___ years, and I will follow the treatment plan specified for me.
- I commit to allow my EAP counselor to communicate with my therapist(s) and authorize them to share information.
- I commit to obtaining a sponsor(s) and attending twelve-step support group meetings, such as Alcoholics Anonymous (AA), Narcotics Anonymous (NA), or Cocaine Anonymous (CA).
- In the event that my work or home situation becomes stressful enough to endanger my sobriety, I commit to immediately inform my sponsor (in AA, NA, CA, etc.), my EAP counselor, my medical department or my supervisor so that appropriate measures can be taken.
- I commit to maintain regular personal contact with my EAP counselor.

My EAP counselor is authorized to contact the following family members or significant others who will be involved in my Recovery Program:

Name(s)   *Telephone Number(s)*

_____   _____

_____   _____

_____   _____

I commit to total abstinence; however, should relapse occur, I will immediately inform my sponsor and attend AA/NC/CA meetings on a daily basis until my EAP counselor determines otherwise. I will also immediately inform my EAP counselor, medical department, and/or supervisor. I understand that this contract may be revised depending on the severity of my relapse.

FIGURE 7-10. Sample recovery contract.

If I am unable to attend recovery activities because of illness, I will, before the appointment time and upon request, provide a written physician's explanation of the medical problem causing the absence.

I understand that failure to comply with the above requirements constitutes a breach of this contract and will be reported to my employer.

I, _____, authorize _____,
     (Name of Client)   (EAP Recovery Counselor)

and _____ to exchange information pertaining to my
     Supervisor

ongoing participation in this recovery contract for the purpose of recovery planning. This authorization shall become effective _____ for a ___ year period, and is subject to revocation in writing by the undersigned at any time.

_____
(Specify date or circumstances under which consent will expire.)

I understand that this information will be used for the purposes noted above and will not be disclosed to any other person or agency without my written permission, except where required by law. I understand that I will receive a copy of this authorization.

_____   _____
Client's Signature   Date

_____   _____
Employer Representative's Signature   Date

_____   _____
EAP Counselor's Signature   Date

FIGURE 7-10. Continued.

The EAP and EAP Counselor are challenged to designing a Testing Program within the framework of an employee's Recovery and Aftercare Plan. Union and/or Company management and the responsible medical representative would help design the plan and support it strongly in order to give leverage to the EAP counselor working with the recovering person. An Aftercare Testing Program should include the following principles:

1. Testing is designed to meet the needs of the client (employee). (Drug(s) of abuse, work schedule, age, sex, cultural background can impact this decision.)
2. Testing should be random.
3. Testing procedures should be reliable and confidential.
4. Testing should be recognized as supportive to the employee's recovery, not punitive.
5. If an employee relapses, the aftercare program should be started anew.

Suggested Time Frame for Testing Recovering Clients (Employees)

Drug of Choice	Months 1-3	Months 4-6	Months 7-9
Cocaine	5 tests per month	4 tests per month	4 tests within each 45 days
Alcohol & Other Water-Soluble Drugs	4 tests per month	3 tests per month	2 tests per month
Marijuana & Other Fat-Soluble Drugs	3 tests per month	2 tests per month	2 tests per month

Drug of Choice	Months 9-12	Months 13-24	Months 24-60
Cocaine	2 tests per month	1 test per month	1 test within each 45 days
Alcohol & Other Water-Soluble Drugs	2 tests within each 45 days	1 test per month	1 test within each 45 days
Marijuana & Other Fat-Soluble Drugs	2 tests within each 45 days	1 test per month	1 test per month

FIGURE 7-11. Employee assistance program testing guidelines and scheduling time frames. (Courtesy of F. Peter Szafran, ACSW, CSW-ACP, LPC, CADAC, a colleague and specialist who manages Aftercare Programs for Corporations as an employee of Human Affairs International, Inc. The author gratefully acknowledges his wisdom and material support.)

Many employers and unions are reviewing their health-care plans and assessing how the EAP can more effectively serve as a "gatekeeper" to expenditures. A Brown & Root Construction Company cost-benefit analysis found that the company spent nearly 100 percent more for mental health and substance abuse care in Houston, Texas, than at other company sites. It was determined that a major problem was that the EAP had little control over mental health benefits. As a result, the Brown & Root EAP now serves as a "gatekeeper" for all of the 34,000 employees nationwide. Brown & Root asked its EAP to select "preferred providers" of counseling and rehabilitation services, and then offered its employees a financial incentive—complete, rather than partial benefit coverage—for using preferred providers. First-quarter 1990 results demonstrated a $590,000 savings to the benefit program.[11]

General Motors, CONOCO, General Electric, IBM, Exxon, and AT&T are a few of the many employers that have been conducting cost-benefit analyses of their EAPs and managed health-care providers. As this research continues and grows more sophisticated, EAP providers will learn in which areas they can become more cost-effective.

SELECTION GUIDELINES

When evaluating outside EAP consulting firms, employers should consider the following elements: (1) staff qualifications; (2) accessibility; (3) services; (4) confidentiality; (5) quality assurance procedures; and (6) educational services. These are the most significant elements of an effective EAP.

Staff Qualifications

Experienced counselors and account manager
Clinical supervision and quality assurance
Minimum of a master's degree in the counseling field or the equivalent
State licensure
Certification as an EAP professional
Assurance of annual continuing education of counselors

Program Accessibility

Network of counselors (full-time or part-time offices) that cover all company/union locations
Around-the-clock emergency availability
Professionally managed and dedicated toll-free line

Flexible scheduling
Ability to manage international incidents

Program Services

Assessment, short-term counseling, and referral capabilities
Substance abuse referral and follow-up reliability
Critical incident (workplace violence, etc.) capability
Conflict resolution, organizational development, and other management consultation capabilities
EAP's understanding of corporate philosophy, culture, subculture, benefit policies, and cultural diversity issues

Confidentiality

Threat of violence or other crisis (e.g., suicidal ideating) procedures
Counseling locations away from work locations
Limited waiting time in reception room
Offices with separate entrance and exit doors, so that clients from the same company/union do not meet as they arrive, wait, or leave

Quality Assurance

Local, regional, and overall clinical supervision
Anonymous client questionnaires
Semiannual (quarterly) utilization reports and annual executive reports
Dedicated account manager
Demonstrated data management abilities
Receptionist and telephone courtesy

Educational Services

Assurance that all supervisors will be offered training within the first year of service
Biannual retraining
Employee orientation
Employee development trainings (stress management, family and drugs, financial management, parenting, smoking cessation)

When an employer/union is searching for an EAP provider, some comparative research is important:

- Ask for references (locally, as well as around the nation).
- Ask for a resume of each counselor who will serve the employees and family members.
- Insist on knowing how many years of service time each counselor has worked with the EAP.
- Do comparative pricing. What do other employers pay per employee per month?
- How much training is given to receptionists, answering services, and the toll-free telephone line staff?
- How many contracts does the EAP currently serve?
- What is the employee/counselor ratio? (Average: 4,000 to 1)
- Does the EAP also manage benefit programs (managed health-care programs)?
- What do other employers say about the EAP? Does the EAP "walk the extra mile" with clients? Are phone calls answered the same day?

SUMMARY

EAPs can boost a firm's productivity and potentially reduce certain costs. All EAPs have in common the philosophy that employees who struggle with difficult personal problems cannot perform effectively, and consequently are less effective workers. An EAP counselor attempts to identify and resolve these problems at the earliest possible moment before mental or physical health has been affected or work performance has been impaired.

The Drug-Free Workplace Act has provided a unique momentum to the EAP field. Each employer or union has the choice of providing only "lip service" or selecting an EAP that will prove effective and productive. In the first instance, the program will be recognized by the employees for what it is "intended to be and do." If a "paper" or "sham" program is set up, there will be minimal utilization and the effect will be minimal. If a good program is initiated, employees will use the EAP, workplace emotional health and morale will improve, and productivity will increase. An EAP program can potentially pay for itself and provide a substantial savings through reduced absenteeism, improved morale, decreased accident rates, and reduced disability payments.

References
1. Reagan, R. Executive Order 12564: Drug-free federal workplace. *Fed. Reg.* 1986;51(Sept. 27):32,889–32,893.
2. National Institute on Drug Abuse. *Model Plan for a Comprehensive Drug Free Workplace Program.* Rockville, Md.: U.S. DHHS publication no. (ADM) 89-1635; 1989.

3. Howard J. Kauffman, president of Exxon Corporation, addressing senior management, supervisory training film. April, 1985.
4. Drug-Free Workplace Act of 1988. Section 5152, pub. law 100–690, 100th Congress, November 18, 1988.
5. Lee, R. The evolution of managed mental health care. *Compensation & Benefits Management* 1988;5:61–66.
6. *Employee Assistance Program Management Letter* 1991;4:1.
7. Francek, J. L., et al. *The Human Resources Management Handbook.* New York: Praegere Special Studies; 1985.
8. Kertesz, L. McDonnell Douglas EAP trims health care costs. *Business Insurance* 1989;23:3–4.
9. Cooke, R. Low-cost help for troubled employees. *Credit Union Management* 1991;14:50.
10. Smith, D. C., and J. J. Mahoney. *EAP Financial Offset Study.* Bridgeton, Mo.: McDonnell Douglas and Alexander & Alexander; 1989:1–18.
11. Bryant, M. Ways to make EAPs more cost effective. *Business & Health* 1990;8:12–20.

8
Monitoring Laboratory Performance

Dennis Crouch, M.B.A., and
Yale H. Caplan, Ph.D.

Historically, drug testing has been performed for a variety of indications, such as poisonings, death investigations, and accident investigations. Under these indications, the patient's medical condition, autopsy findings, or an accident investigation report detail the circumstances surrounding the event. These data provide evidence in addition to laboratory test results from which medical treatments can be instituted, legal issues argued, or crime scenes reconstructed. This supportive evidence provides a holistic approach to the case and improves confidence in laboratory test results.[1] The current practice of workplace drug testing represents a comparatively unique situation where the laboratory's results of drug presence may be the single piece of evidence on which legal or disciplinary actions are taken.[2] The laboratory is located apart from the urine collection site, the identity of the donor is unknown, and no case or clinical history is provided to the laboratory. The laboratory and employer may rely entirely on the analytical findings. Disciplinary actions are taken primarily on the basis of the laboratory results. For this reason, laboratory analyses must be accurate, reliable, and legally defensible.

The results of performance (or proficiency) testing surveys have raised concerns about the accuracy of drug testing laboratories. A 1985 publication on the Centers for Disease Control (CDC) blind testing study exemplifies the severity of the problem. The CDC found false-positive rates in excess of 60 percent and error rates up to 100 percent in some drug classes when the laboratories tested for barbiturates, amphetamines, methadone, cocaine, and codeine.[3] A study conducted from 1986 to 1987 found a 31 per-

TABLE 8-1. False-Negative and False-Positive Performance Testing Findings

Drug/Class	False Negative (%)	False Positive (%)
Amphetamine	4.9	1.6
Benzoylecgonine	3.3	2.9
Morphine	2.3	3.3
Codeine	0.7	0.6
THC-COOH	10.5	0.7
Phencyclidine	8.6	0.4

Source: Data from Standefer.[6]

cent false-negative rate on blind performance testing samples and a 17 percent false-negative rate on known performance testing samples.[2]

Other surveys have demonstrated better performance by laboratories. Accuracy rates of 98.6 percent for opiates, 100 percent for cannabinoids, 98.7 percent for amphetamines, 99.2 percent for cocaine, and 99.5 percent for phencyclidine were achieved in a 1987 study.[4] Another 1987 study found a 92 percent accuracy rate on blind performance testing samples submitted to selected laboratories.[5] By contrast, a 1990 study found relatively high false-positive and false-negative rates, as shown in Table 8-1.[6] Given the consequences of employee drug tests, critics may argue that the only acceptable performance is 100 percent accuracy.

The questionable accuracy of laboratories in performance testing programs and the increasing demand for testing of specimens collected in urine drug testing programs led to predictions that regulatory standards for laboratories would be established.[7,8] The 1988 Mandatory Guidelines for Federal Workplace Drug Testing Programs (NIDA [National Institute on Drug Abuse] Guidelines) contain detailed requirements that laboratories must fulfill to achieve and maintain NIDA certification.[9] The College of American Pathologists (CAP) and the American Association of Clinical Chemistry (AACC) have jointly established a certification program for forensic urine drug testing, with its own set of detailed requirements.[10] Both the NIDA and CAP/AACC programs require that laboratories have internal quality control and quality assurance programs, open performance testing, and on-site inspections to achieve and maintain certification. Quality control and quality assurance are the means by which a laboratory internally monitors its performance. Performance testing, on-site inspections, and external blind performance testing are the methods by which external entities can monitor a laboratory's performance.[11,12] A brief description of each monitoring technique follows:

- *Quality control (QC)*. Laboratory accrediting bodies provide guidelines under which laboratories establish their internal QC procedures. These procedures are individualized to fit each laboratory's operation, equipment, process flow, and personnel. QC samples are analyzed concurrently with donor urine specimens. If the QC sample results fall within predetermined values, the analysis batch is accepted; if not, the batch is rejected and the analyses must be repeated.[13] Therefore, quality control samples and QC sample acceptance rules are extremely important in establishing the accuracy and reliability of screening and confirmation data. QC results should be recorded at the time of analysis and be available for review by the laboratory and by outside parties.

- *Quality assurance (QA)*. QA is the review of all procedures and operations of the laboratory to ensure optimal operations. QA includes review of QC, review of analytical records, staff training programs, and instrument maintenance. QA also includes monitoring of water quality, calibration of balances, stability of electrical power sources, pipet calibration, the accuracy of thermometers, and other less obvious laboratory operations.[14] QA serves such an essential function that a laboratory may invest up to 15 percent of its total man-hours in QA.[8]

- *Performance testing (PT)*. Certifying agencies and organizations prepare and challenge the laboratory with PT samples. These samples contain predetermined concentrations of drugs. The laboratory tests the samples, submits its results to the certifying organization, and is graded on its ability to detect, quantify, and report results correctly. Grading is weighted, with an emphasis on detecting the proper drug(s) and accurately quantifying the concentration of each drug or metabolite present. A satisfactory score is required to achieve and maintain accreditation. Failure to meet the minimum passing score on a single batch of performance testing samples usually results in the laboratory being required to identify and correct the source(s) of the error. False positive reports, false negative reports, or continued poor performance may result in probation, suspension, or revocation of certification.

- *External blind performance testing*. External blind performance testing samples are specially prepared samples that employers submit to the laboratory, which processes them as if they were donor specimens. These samples should be distinguished from internal blind control samples, which are required for accreditation by both NIDA and CAP/AACC, but are inserted into the testing process by the laboratory (see the QC section below).[9,15] The laboratory has no knowledge of the identification or drug content of external blind samples. For agencies testing federal employees, current federal regulations require that a minimum of 10 percent of all specimens sent to the laboratory be external blind samples. Private-sector programs subject to either the Department of Transportation (DOT) or Nuclear Regulatory Commission (NRC) regulations must submit at least 3 percent of all specimens as blind samples.[16,17] Since the Medical Review Officer (MRO) often acts as "gatekeeper" of results for the employer, monitoring and managing the external blind performance testing program may be part of the MRO's responsibilities.

- *Laboratory inspections.* Federal laboratory accreditation bodies and many state and local agencies use on-site inspections as an integral part of their certifying programs.[18] These inspections are comprehensive. Inspection teams ask detailed questions about the laboratory's procedure manual, specimen receiving process, chain of custody (COC) documentation, records and record security, personnel competency, reagent preparation, quality control practices, reporting protocols, equipment and equipment maintenance, screening and confirmation methods, and safety precautions.[15,19] A thorough on-site laboratory inspection requires a trained professional with experience in analytical and forensic toxicology.

INTERNAL PERFORMANCE MONITORING
Quality Control

Laboratories insert QC samples in each screening and confirmation test batch. Results from these samples are used as a criterion to determine the acceptability of results for donor specimens. Three aspects of laboratory QC are calibrators (or standards), open controls, and internal blind controls. Laboratories have the option of purchasing QC materials or preparing them in-house. When calibrators and controls are prepared in-house, urine is collected, pooled, and tested to verify that it is drug-free. The drug-free urine is then fortified with known amounts of the drug(s) or metabolite(s) to attain the desired concentration(s). The concentrations are then analytically verified by the laboratory.

Calibrators

Calibrators are the QC material used by the laboratory to establish the response of the analytic instrumentation. Screening calibrators are usually purchased directly from immunoassay test kit manufacturers and are used to establish the instrument response at the screening cutoff concentration. The response of donor specimens can be compared to the calibrator response to distinguish negative from presumptive positive specimens. Gas chromatography/mass spectrometry (GC/MS) calibrators are purchased or prepared by the laboratory. Drug-free urine is fortified with drug/metabolite at increasing concentrations to establish a standard curve. This standard curve is derived from the GC/MS response signal to the calibrators. Since the calibration curve becomes the reference to which donor specimens are compared and quantified, the laboratory must have confidence in the reference drug material from which the calibrators are prepared and in the validity of the calibration curve.

Open Controls

Open controls are known QC samples that are treated exactly like donor urine specimens throughout the analytic process. The instrument response on controls is compared to calibrator responses. Successful qualitative and quantitative performance on open controls is a criterion for acceptance of the testing batch and reporting of the donor results. To ensure accuracy, the reference materials used to prepare controls should be obtained from a different source than those used to prepare calibrators. If this is not practical, separate weighings of the reference material for calibrator and control preparation are suggested.[20] Calibrators and controls made from the same weighing of a reference material may test analytical precision, but not accuracy.

The use of controls has its scientific basis in probability theory.[21] The probability of not detecting analytical data errors is balanced against the probability of falsely rejecting accurate analytical data.[22] From this statistical basis, Westgard developed the control acceptance rules shown in Figure 8-1. Westgard's rules assume that a wealth of historical QC data are available in the laboratory and allow for acceptance of data while further testing proceeds. These rules have been widely accepted by clinical laboratories, but are difficult to apply to forensic testing due to the need for quick turnaround of definitive results. Most forensic laboratories have developed simpler rules for accepting controls and test batches, for example, controls must be within 20 percent or 2 standard deviations of their validated concentrations.

The NIDA Guidelines require that each screening batch contain a certified drug-free sample, calibrators, a positive control within 25 percent of the cutoff concentration, at least one internal blind control, and quality control samples constituting at least 10 percent of the batch size. Each GC/MS confirmation batch must contain a certified drug-free sample, calibrators, and a positive control within 25 percent of the cutoff concentration. Control values from GC/MS confirmations should be recorded in a standardized format that allows for observation of trends and rapid detection of outlying values.[22]

1. If one control is greater than 2 standard deviations from the expected value, this is a warning, and additional control data should be evaluated.
2. If one control is greater than 3 standard deviations from the expected value, the data should be rejected.
3. If two consecutive controls are greater than +2 standard deviations or greater than −2 standard deviations from the expected value, the data should be rejected.
4. If, within a run, one control is greater than +2 standard deviations and one is greater than −2 standard deviations from their expected values, the data should be rejected.
5. If, in consecutive analyses, one control value exceeds +1 standard deviation or −1 standard deviation from the expected value, the data should be rejected.
6. If, in 10 consecutive analyses, one control value falls consistently on one side of the mean, a systematic error is present.

FIGURE 8-1. Westgard quality control rules.
Source: Based on Westgard and Barry.[22]

Internal Blind Control Samples

Most certification programs require the analyses of internal blind control samples with each batch of specimens screened.[9,10,23] Internal blind control samples should constitute at least 1 percent of the total number of samples tested in a batch, with at least one per batch. These samples may be purchased, pooled from previously tested specimens, or prepared in-house. To be truly blind to the technician doing the testing, the samples should be indistinguishable from donor specimens. The content and the concentration of drug or metabolite in the samples may vary. Many laboratories have adopted a policy of analyzing 80 percent drug-free samples and 20 percent containing one or more of the drugs or metabolites being analyzed. Neither NIDA or CAP/AACC require internal blind control sample challenges of GC/MS testing; however, when included they provide an added level of confidence.

The design of the internal blind program is left to the discretion of the laboratory. Internal blind samples may be:

1. Inserted into the testing batch during accessioning, and undergo screening analysis, only. This serves as a check on the screening procedures.
2. Inserted into the batch during accessioning, and undergo screening and

confirmatory analyses. This serves as a check on both screening and confirmatory procedures.
3. Submitted to the laboratory through actual or fictitious client accounts, and decoded after screening (or after confirmatory) testing. This serves as a check on accessioning, as well as screening (and confirmatory) testing.

In the third scenario given, both accessioning and analytical procedures are challenged.

In an effective program, the internal blind sample results are a primary criterion for acceptance of analytical results. Screening data should be accepted and reported only after the certifying scientist or the quality control officer has reviewed the data, identified the internal blind samples, and ensured that the sample results are consistent with their known drug content(s). The laboratory must maintain documentation of this process.

Quality control serves to document analytic procedures. But even competent laboratories make errors. Quality control should help identify errors so that they can be corrected before results are reported. QC standards for data acceptance should be rigorous because workplace drug testing results often have legal implications and disciplinary consequences. A sound QC program instills laboratory self-confidence and improves the reliability of test results.

Quality Assurance

QA is the umbrella of procedures used by the laboratory to ensure that all processes are operating optimally. It provides internal laboratory oversight designed to enhance testing quality and reliability. Federal regulations define quality assurance as a program, "including but not limited to specimen acquisition, security and reporting of results, initial and confirmatory testing and validation of analytical procedures. Quality assurance shall be designed, implemented and reviewed to monitor each step of the process of testing for drugs."[9,16,17] A well-designed QA program monitors all aspects of the laboratory, focuses on achieving accurate and reliable results, and instills an awareness among the laboratory personnel of the need for continuous evaluation of the laboratory's operations.

QA review of specimen handling is essential. Accessioning personnel and analysts check and recheck the identity of all specimens at the time of receipt and throughout screening and confirmation testing. Specimen and specimen aliquot security require well-designed physical systems and constant QA review by supervisory staff to ensure that all locks are secured, all keys are accounted for, and unauthorized personnel are not admitted to secure areas. If constant attention is not placed on the security of specimens, the legal and scientific credibility of forensic test results may be jeopardized.

Certain QA functions pertain to personnel and personnel training. The laboratory director must:[20]

1. Ensure that the staff is properly trained and has appropriate experience;
2. Maintain an adequate number of staff;
3. Review and approve all standard operating procedure (SOP) manual changes;
4. Implement a QA program;
5. Establish QC criteria;
6. Ensure that testing methods are accurate and reliable;
7. Provide appropriate responses to remedial and corrective action requests;
8. Review reports.

QC checks ensure that all calibrators, controls, and testing reagents are properly prepared. Data related to the preparation, validation, and use of these materials must be maintained. Periodic review of these records is a QA function. A typical challenge to a laboratory during an inspection or during litigation is to ask the laboratory staff to: (1) identify the lot number of reagents, controls, and calibrators used during the analysis of the specimen in question; and (2) provide the accompanying validation records. Effective QA procedures are needed to ensure that legally defensible documentation can be produced.

Traditionally, QA includes maintenance of equipment. For radioimmunoassay testing, maintenance records should include radiation background checks, gamma counter well checks, and gamma counter well calibration. Enzyme immunoassay instrumentation may require daily spectrophotometer checks, temperature zone monitoring, and pipet volume calibrations. GC/MS instrumentation requires extensive daily maintenance. Tuning is an operation designed to optimize the MS, and is performed in conjunction with calibration (mass assignment). QA data from daily tuning and calibration provide a historical record from which assessment can be made about instrument function and the need for service. Chromatography parameters must also be evaluated daily. Deviations in temperature-dependent zones will adversely affect chromatographic separation of the tested analytes. QA procedures ensure that daily maintenance operations are performed, instrument modifications are recorded, and maintenance records are reviewed.

The list of laboratory equipment requiring QA monitoring to ensure optimum performance is extensive. For example, the NIDA Guidelines require that balances and pipets be calibrated to ensure analytical accuracy; glassware, water sources, and centrifuges be evaluated to avoid contamination problems; constant temperatures be maintained in specimen storage, reagent storage, and sample preparation facilities to avoid degradation; power sources be monitored to ensure optimum instrument life and operation; and fume hoods, waste disposal, and safety equipment be periodically checked to ensure a safe working environment.

In summary, QA is a comprehensive preventive maintenance program. It should be designed to assist the laboratory in avoiding errors and in detecting the source of errors if they occur. When conscientiously incorporated into the laboratory, QA provides historical records that indicate all instruments, QC materials, and reagents were reliable at the time of testing.

EXTERNAL PERFORMANCE MONITORING

Open Performance Testing

In PT, urine specimens fortified with drugs or drug metabolites are submitted to monitor the laboratory's analytical accuracy and precision. In open performance testing, these samples are identified to the laboratory, which understands that it will be graded on its accuracy in testing them. Open PT should, therefore, represent a laboratory's best performance. NIDA, CAP/AACC, DOD,[23] the American Association of Bioanalysts,* the California Association of Toxicologists, the New York State Department of Health, and many other groups prepare and challenge laboratories with PT samples. The fortified PT sample concentrations are verified at reference testing laboratories prior to submission to the laboratories. After analysis, each laboratory forwards its results to the submitting organization for scoring.

The NIDA Guidelines require open performance testing in both the application and continuing phases of certification. The PT samples are prepared at drug/metabolite concentrations at least 20 percent greater than the respective immunoassay cutoffs for screening and the GC/MS cut-offs for confirmation.[9] CAP/AACC utilizes slightly different criteria and prepares the samples near a "minimum threshold of detection" concentration.

For both the NIDA and the CAP/AACC programs, certification is contingent on successfully completing three consecutive open PT challenges and passing an on-site inspection.[9,10] Applicant laboratories in the NIDA program receive batches of 20 PT samples that contain drug-free samples and samples that contain one or two target drugs or metabolites. The 20

---

*Contact the American Association of Bioanalysts, Proficiency Testing Service, 205 West Levee, Brownsville, TX 78520-5596; (512) 546-5315.

samples are processed through the accessioning, screening, confirmation, and reporting procedures as if they were donor specimens. To evaluate quantitative accuracy and precision at concentrations below the screening cutoff values, the laboratory is directed to test certain PT samples by GC/MS regardless of the screening result. Successful completion of two consecutive batches of PT samples qualifies the laboratory for an on-site inspection and a third PT challenge. Minimum acceptable PT scores for applicant laboratories are as follows:[9]

1. At least 90 percent qualitative accuracy cumulative for the three batches;
2. At least 80 percent of the quantitative values reported must be within +20 percent or −20 percent, or +2 SD or −2 SD of the predetermined reference laboratory mean (whichever is greater);
3. No quantitative value can differ by more than 50 percent from the reference group mean;
4. At least 50 percent of the challenges for each drug must be successfully detected and quantified;
5. No false positives can be reported.

CAP/AACC scores each survey by the following formula:[10]

Each positive sample correctly identified	4 points
Each false negative	0 points
Each false positive	−25 points*
Each quantitation within 1 SD of participant mean	2 points
Each quantitations within 2 SD of participant mean	1 point
Each negative sample correctly reported	4 points

Criteria for acceptable performance on open PT challenges are essentially the same for applicant and certified laboratories. Unacceptable performance by certified laboratories on a single performance testing batch does not automatically mean decertification. The laboratory may instead be required to explain unacceptable results and take corrective actions designed to improve future performance. There are variations in the scoring of PT samples. For example, quantitative scoring may be based on a mean concentration obtained by participants or by reference laboratories.[9,10,12] Scoring changes are anticipated in the NIDA program. One 50-percent

---

*Results in failure of the PT challenge; 80 is the minimum passing score.

quantitative error may be tolerated per three PT batches in certified laboratories and one 50-percent quantitative error overall may be allowed for applicant laboratories.[12]

There are also variations in the frequency of PT challenge submissions from certifying groups:

1. CAP/AACC sends (and NIDA has received a recommendation to send) PT samples quarterly.[10,12]
2. NIDA sends to laboratories six batches per year on alternate months.
3. DOD sends monthly challenges to the laboratory.[23]

An open PT report to the laboratory contains reference laboratory results, participant laboratory results, and discussions of problem samples. The report allows the laboratory director to compare his or her laboratory's qualitative immunoassay accuracy and quantitative GC/MS accuracy to that of other laboratories. Trends indicating declining performance can be identified and corrective action can be taken before donor testing accuracy is jeopardized. Qualitative (false positive or false negative) inaccuracies on PT samples are symptomatic of major laboratory problems. These problems may be linked to poor data review and reporting protocols, misassignment of screening or confirmation cutoffs, inability of the screening or confirmation method to discriminate near the cutoff, inadequate laboratory methods, or other explanations.

Quantitative values deviating more than 20 percent or 2 standard deviations from the accepted mean are cause for alarm and require an immediate corrective response. Consistent values more than 10 percent or 1 standard deviation from the expected values are a warning that a systematic error exists in the laboratory's analytic procedures for that drug/metabolite. Widely scattered values at or near the tolerance limits for a drug indicate that the analytical method lacks precision. A trend toward concentrations consistently lower, or consistently higher, than the group mean may indicate a bias in the calibrators for the analysis. A laboratory should not be expected to score 100 percent on all open PT challenges. Random error in any analytical system produces some deviation. The conscientious laboratory uses PT results to modify and improve its operation and to reduce the potential for significant error.

Open PT results may be used by inspection teams, clients, MROs, and other interested parties to assess the accuracy and reliability of the laboratory. PT performance also provides an interlaboratory data base from which the laboratory can be judged. It provides the laboratory with a benchmark for comparison, problem identification, and problem solving.

Good PT scores should instill confidence in clients that the laboratory's tests are accurate and reliable.

External Blind Performance Testing

External blind performance samples provide the best indicator of day-to-day laboratory performance. External blind samples are subject to GC/MS confirmation; by comparison, internal blind controls may not be. Because the laboratory cannot identify external blind samples, the analyses better reflect performance on routine samples.[3] External blind performance testing programs help serve to[24]:

1. Ensure accurate test results;
2. Detect laboratory testing problems;
3. Instill employee confidence in the program;
4. Demonstrate testing reliability if there are legal challenges to the program.

The federal drug testing regulations require that covered employers have external blind performance testing programs. External blind testing is an effective monitoring tool that should be considered for every program, even where federal requirements do not apply.[5] The NIDA Guidelines establish the following rules for the submission of external blind performance testing samples:

1. For the first 90 days of testing at a laboratory, at least 50 percent of the samples, up to 500 in total, must be external blind samples;
2. Thereafter, a minimum of 10 percent, up to 250 samples per quarter, must be external blind samples;
3. 80 percent of the external blind samples should be negative;
4. 20 percent of the samples should contain one or more of the drugs tested;
5. Positive samples should be distributed in approximately equal proportions to the five drugs being tested;
6. The cost of external blind samples should be born by the employer.

Table 8-2 summarizes requirements regarding external blind PT.

As with many aspects of workplace drug testing, production of external blind performance testing samples is a new industry and regulations under which these materials must be manufactured, labeled, and marketed are not clearly established at the time of this writing. Food and Drug Administration (FDA) regulations and Drug Enforcement Administration (DEA) reg-

TABLE 8-2 Requirements for External Blind Performance Testing

Topic/Agency	NIDA	DOT	NRC
Submission rate			
First 90 days	50%	3%	50%
MAX/QTR	500	N/A	500
Thereafter MAX/QTR	250	100	250
% Negative	80	80	80
% Positive	20	20[a]	20
Distribution of positives by drug	even	even	even
Discrepancies, General			
Investigation required	+	+	+
Report from lab required	+	+	+
Corrective action plan required	+	+	+
Response to false-positive results			
Investigation required	+	+	+
Who investigates?	Secretary DHHS	DOT or DHHS	Licensee
Report from lab required	+	+	+
Corrective action plan required	+	+	+
QC material required[b]	+	+	+
Test to last successful PT[c]	+	+	+
On-site inspection	Optional	Optional	Optional
Positives retested in 2nd lab	Not precluded	Not precluded	Optional
Report to DHHS	N/A	Optional	Required

[a] Employers with less than 2,000 employees may submit negative samples or split samples from nonregulated employees.
[b] QC material from the failed batch must be provided upon request of the agency.
[c] The laboratory must retest all reported positives (for the drug in question), from the time of the last successful PT, to the time of the false-positive report.[11,20,21]

ulations may apply to the manufacture and distribution of the samples.[25,26] NIDA maintains a list of vendors of blind performance testing samples and providers of performance testing services.* This list of vendors is provided as a service; NIDA does not endorse any vendors. Laboratories, MROs, and employers currently involved in workplace testing may also be reliable sources for referrals to vendors.

There are no explicit requirements for the preparation, certification, validation, encoding, or shipment of PT samples. The following is a typical production scenario: Urine is collected and pooled from volunteers or paid subjects. Depending on the size of the production operation, pooled

---

*This list can be obtained by contacting NIDA's Division of Applied Research, 5600 Fishers Lane, Rockville, Md. 20857.

batches may contain several gallons of urine. The pooled urine batch is analyzed by immunoassay and GC/MS to ensure that it contains none of the drugs or metabolites to be tested for in the employee testing program. The pool is centrifuged or filtered to remove precipitates and the pH is adjusted to the normal physiological range. Preweighed amounts of drugs/metabolites are dissolved in a solvent and added to the pooled urine to attain the desired drug concentration(s). The fortified urine is then thoroughly mixed. Aliquots are removed and analyzed. If the pool does not contain the desired concentration(s), additional analyte or urine is added, and the analysis is repeated. Once validation demonstrates that the desired drug/metabolite concentration has been achieved, sample preservatives such as sodium fluoride may be added. The fortified urine pool is aliquoted into individual specimen bottles. Each bottle must contain the minimum volume required by the NIDA Guidelines (60 ml as of 1991, though this is anticipated to drop to 30 ml). Samples should be stored and shipped frozen to extend their shelf life.

Validation of the drug concentration is the most critical step in PT sample production. Failure to properly validate that the fortified urine pool contains the intended drug/metabolite concentration may result in a sample with less analyte than expected, which may lead to an apparent false-negative report. Some vendors spike the drugs/metabolites at concentrations very near the NIDA cutoffs, with the intent of maximizing the challenge to the laboratory. PT samples with concentrations close to the cutoffs may produce false-negative results because drugs and metabolites may degrade over time, and because of normal analytic variability. NIDA and AACC recommend that analyte concentrations in the external blind performance samples be at least 125 percent and 120 percent of the cut-offs respectively.[12,24]

Figure 8-2 presents typical costs for external blind samples and services. Prices and services vary considerably. Reputable providers should:

1. Provide references of current customers;
2. Demonstrate an understanding of the current federal program requirements for the samples;
3. Be aware of FDA, DEA, and other regulations;
4. Provide suggestions for encoding the samples, so that they are blinded to the laboratory;
5. Provide records of urine certification and concentration validation, if needed;
6. Understand the NIDA Guidelines and technical issues of laboratory certification;

> ### External Blind Performance Testing Vendors
>
> Average price per negative $17.70
> Average price per positive $34.25
> Reviewing discrepancies   No charge
> On-site inspection of a laboratory $131.25/hour + expenses
> Do you manage test results?[a]   2 Yes and 3 No
> Do you manage programs?[b]   2 Yes and 3 No
>
> What do you do if there is a discrepancy?[c]
>
> Review external blind validation data (4).
> Request sample analysis data from the laboratory (3).
> Reanalyze the sample (3).
> Respond as the client requests (3).
>
> [a]Does the vendor offer the service of receiving and tracking laboratory results?
> [b]Does the vendor offer the service of receiving, tracking and reporting laboratory results to employers, MROs, and laboratories?
> [c]This question was posed open-ended. In parentheses are the number of vendors who had each response.

FIGURE 8-2. Summary of prices and services from external blind performance testing vendors. (Based on the authors' telephone interviews with five randomly selected vendors.)

7. Assist in investigating and reconciling discrepancies when expected results are not obtained;
8. Ensure the purchaser that conflict(s) of interest do not exist through financial arrangements with multiple clients, laboratories, the National Laboratory Certification Program (NLCP), CAP/AACC, or colleagues.

The collection sites should be given instructions for encoding of the external blind samples. Encoding should serve to disguise the samples so that the laboratory can not distinguish them from actual donor urine specimens. Therefore, external blind samples should be submitted with (mock) COC documentation, should be appropriately labeled, and should be able to pass the laboratory's integrity and adulteration tests. Laboratories may test hundreds to thousands of specimens daily and encounter many external blind samples. They may readily identify external blind samples encoded with sequential social security numbers or atypical employee numbers. The collection site should avoid these and other patterns that can "tip" the laboratory that a sample is an external blind sample. Examples of other such patterns include the presence of the parent drug without its corresponding metabolite, multiple drugs in the sample, combinations of drugs not ordinarily encountered in real samples, repetitively submitting the same sample, repetitively submitting samples that contain drugs that are rarely encountered such as phencyclidine, and the presence of organic solvents (used during PT sample production).

Because the MRO receives all test results from the laboratory (in federally mandated testing programs) the MRO is in a good position for decoding, tracking, and evaluating the laboratory's external blind performance testing sample results. (Table 8-3 provides an example of how PT samples and results can be tracked.)

When a blind performance testing result discrepancy occurs, the sample must be identified to all parties; the COC and accompanying documents must be reviewed; the accessioning, analytical, and reporting processes must be audited; and the sample preparation and submission should be critically reviewed. External blind PT results should be shared with the employer and the laboratory. In practice, many substance abuse testing programs lack systematic and timely procedures to capture the data and forward them to the laboratory. Some private employers have recognized the liability associated with a failed external blind performance test sample and judiciously monitor their results. However, few groups provide feedback to the laboratory except when a discrepant result occurs.

Each investigation of an external blind PT discrepancy should identify the source of the problem and should result in corrective action to prevent the problem from recurring. There are numerous potential causes for an external blind performance testing sample failure.[24] Common reasons for sample discrepancies are presented in Tables 8-4, 8-5, 8-6, and 8-7. Failures can be traced to:

1. The sample's vendor (Table 8-4);
2. The collection site (Table 8-5);
3. The laboratory (Table 8-6);
4. The recipient of the results (Table 8-7);
5. A break in communications between any of these key entities.

Poor sample quality is a common cause of false-negative external blind PT results. Poor sample quality can be established through failure of a vendor's sample at more than one laboratory, or through a reanalysis of the sample that demonstrates the drug concentration was less than the vendor claimed. Unfortunately, many high volume laboratories discard negative specimens shortly after reporting, precluding reanalyses. Major causes of

TABLE 8-3 External Blind PT Results as of 12/31/91

Sample	Date Sent	Client	Pseudonym	Lab	Panel #	Result (ng/ml)	Drug and Ref. Concentration (ng/ml)	Comments
1A	10/04/91	XYZ	John Blair	ABC	1001	Cocaine 110 Benzoylecgonine 1200	Cocaine 350 Benzoylecgonine 900	Split sample with 1B
1B	10/04/91	XYZ	James Williams	ABC	1001	Cocaine <100 Benzoylecgonine 1200	Cocaine 350 Benzoylecgonine 900	Split sample with 1A
2	10/09/91	XYZ	George Carter	ABC	1001	Drug-free	Drug-free	
3	10/14/91	XYZ	Peter Pajor	ABC	1001	Drug-free	Drug-free	
4A	10/18/91	XYZ	Edward Patterson	ABC	1001	THC 147	THC 120	Split sample with 4B
4B	10/18/91	XYZ	Julie Robertson	ABC	1001	THC 146	THC 120	Split sample with 4A
5	10/23/91	XYZ	Helen Woods	ABC	1001	Drug-free	Drug-free	
6	10/28/91	XYZ	David Woodson	ABC	1001	Drug-free	Drug-free	
7	11/01/91	XYZ	Ralph Arwood	ABC	1001	Drug-free	Drug-free	
8	11/06/91	XYZ	Craig Nelson	ABC	1001	THC 126	THC 120	
9	11/11/91	XYZ	C. H. Choudry	ABC	1001	Drug-free	Drug-free	
10	11/15/91	XYZ	Richard Bachman	ABC	1001	Drug-free	Drug-free	
11	11/20/91	XYZ	Carolyn Violet	ABC	1001	Drug-free	Drug-free	
12	11/25/91	XYZ	T. J. Clark	ABC	1001	THC 130	THC 120	
13	11/27/91	XYZ	Adam Stokes	ABC	1001	Drug-free	Drug-free	
14	12/03/91	XYZ	Marcus Gunn	ABC	1001	Drug-free	Drug-free	
15	12/06/91	XYZ	Max Planck	ABC	1001	Drug-free	Drug-free	
16	12/11/91	XYZ	Henry Winckler	ABC	1001	Drug-free	Drug-free	
17	12/16/91	XYZ	Jonathan Anderson	ABC	1001	Drug-free	Drug-free	
18	12/29/91	XYZ	Deborah Wheaton	ABC	1001	THC 113	THC 115	

TABLE 8-4 Potential External Blind Performance Testing Sample Errors: Vendor

	Result	
Errors	False Negative	False Positive
1. Failure to certify negative urine		***
2. Poor-quality reference "spiking" material	***	
3. Inadequate procedures for concentration verification	***	
4. Prepared at wrong cutoff	***	***
5. Prepared with wrong analytes	***	***
6. Clerical error in encoding the sample	***	***
7. Switching sample A with sample B	***	***
8. Communication error with purchaser	***	***
9. Limited sample shelf life	***	
10. Poor-quality urine (pH or specific gravity) unacceptable	Sample rejection	

TABLE 8-5 Potential External Blind Performance Testing Sample Errors: Collection Site

	Result	
Errors	False Negative	False Positive
1. Clerical error in decoding the sample from the vendor	***	***
2. Clerical error in encoding the sample to the laboratory	***	***
3. Shipping delay	***	***
4. Incorrect requisition (wrong test/wrong laboratory)	***	***
5. Exchanging sample A for sample B	***	***
6. Improper storage of sample	***	***
7. Failure to label	Sample rejection	
8. Recording an out-of-range temperature	Sample rejection	

false-positive and false-negative external blind performance testing sample reports have been clerical and administrative errors. These errors may occur during accessioning, testing, and reporting. Clerical errors may also occur during preparation of the samples by the vendor and at the collection site.

False-positive external blind PT sample errors must always be investigated. DOT allows the option of a Department of Health and Human Services (DHHS) or DOT investigation of such discrepancies and sets no fixed

TABLE 8-6 Potential External Blind Performance Testing Sample: Laboratory Errors

Errors	Result	
	False Negative	False Positive
1. Laboratory accessioning error	***	***
2. Laboratory sample chain of custody error	***	***
3. Laboratory aliquot chain of custody error	***	***
4. Screening aliquoting error	***	***
5. Screening clerical error	***	***
6. Inadequate screening method—quality	***	***
7. Inadequate screening method—data review	***	***
8. Inadequate screening method—QA procedures	***	***
9. Inadequate screening method—assay interference	***	***
10. Inadequate screening method—lack of method validation	***	***
11. GC/MS aliquoting error	***	***
12. GC/MS clerical error	***	***
13. Inadequate GC/MS method—quality control	***	***
14. Inadequate GC/MS method—data review	***	***
15. Inadequate GC/MS method—QA procedures	***	***
16. Inadequate GC/MS method—assay interference	***	***
17. Inadequate GC/MS method—lack of method validation	***	***
18. Laboratory reporting error (internal)	***	***
19. Laboratory computer transmission error	***	***

TABLE 8-7 Potential External Blind Performance Testing Sample Errors: Recipient of Results

Errors	Result	
	False Negative	False Positive
1. Clerical error in decoding the sample from the laboratory	***	***
2. Clerical error in encoding the sample from the vendor or collection site	***	***
3. Shipping results to wrong employer	***	***
4. Employee wrongly identified	***	***
5. Communication error with lab or collection site	***	***
6. Electronic transmission error with employer	***	***
7. Electronic transmission error with laboratory	***	***

investigation time frame.[16] The NRC requires the licensee to perform the investigation and requires the investigation be completed within 30 days.[17] The NIDA Guidelines require the secretary of DHHS to:

1. Investigate any unsatisfactory result;
2. Document the cause of the erroneous result;
3. Notify other agencies of the unsatisfactory external blind performance testing results;
4. Ensure that corrective action is taken to prevent a reoccurrence of the error.

The NIDA Guidelines are unclear in their definition of unsatisfactory performance. The requirement to investigate "any" unsatisfactory result (p. 11,985) is in apparent conflict with the statement, "Performance on [external blind performance testing samples] shall be at the same level as for the open or non-blind performance testing" (p. 11,988).[9] The latter statement implies that the 90 percent qualitative accuracy required for open PT samples is sufficient for external blind samples. This would allow a limited number of false-negative reports.

The NIDA Guidelines distinguish between administrative false-positive errors and technical/methodological false-positive errors. Administrative false positives require corrective action. Methodological errors are considered more serious since they are systems-related and may affect other tested specimens. A methodological error requires retesting of all specimens reported positive for the drug in question back to the last successful performance testing cycle. Under extreme conditions, the Secretary of DHHS may suspend or revoke a laboratory's certification. CAP/AACC suggests that a single false-negative error is not cause for alarm, but repeated errors or false-positive external blind performance testing errors should be investigated.[24]

External blind performance testing is a new concept with many problems and few established regulations. Areas that need to be further addressed include:[12]

1. Methods to minimize cost;
2. Distribution of test results to submitting agencies, laboratories, and certifying agencies;
3. Methods for producing, verifying, and stabilizing external blind performance testing samples.

Laboratory Inspections

On-site inspections are the most revealing means of monitoring laboratory performance. Most accrediting agencies require an on-site inspection as a condition of certification. Employers should reserve the right to inspect the laboratory as part of their contractual arrangement.

Inspections vary in frequency, format, inspection team size, duration, and, consequently, in cost. NIDA inspects each certified laboratory on a semiannual basis, CAP/AACC requires an annual inspection with an interim 6-month laboratory self-evaluation, and the DOD requires quarterly inspections.[9,23] Even where not required by federal regulations, quarterly or semiannual on-site inspections are reasonable.[5]

Inspection protocols and reports may follow a predetermined checklist, use an audit style, or be narrative. The checklist format has the advantage of ensuring that all aspects of the laboratory are evaluated, and that the inspections are objective and consistent.[19,20] Both the NIDA and the CAP/AACC checklists contain questions that are critical to passing the inspection. Certification cannot be issued, or continued, if the laboratory fails these critical questions. The current CAP/AACC inspection checklist has over 280 questions of which 27 are critical.[15] The NIDA inspection checklist contains over 360 questions of which 79 are critical.[19] Table 8-8 shows the distribution of questions, and of critical questions, between the inspected laboratory sections. This distribution reflects the relative importance that NIDA and CAP/AACC place on each area. Other programs, such as NRC, use an audit inspection format.[27] Private-sector inspections are usually less structured and are reported in narrative form.

The number of inspectors per inspection team and the duration of the inspection vary between the programs. DOD inspection team size varies by armed service, but may include an attorney to audit records for legal sufficiency, a forensic toxicologist to perform the basic laboratory inspection, and a DOD program representative to address policy issues. The inspections last 2 or more days based on the testing volume of the laboratory.[23] NIDA uses teams of three inspectors and a 2-day inspection.[20] CAP/AACC varies the inspection team size with the number of specimens tested daily by the laboratory. Usually, teams have at least two inspectors and inspections last 1 day.[10] Inspections by state and local agencies and private employers are usually conducted by one or two inspectors and last for 1-2 days. Larger inspection teams offer the advantages of multiple perspectives, dilution of individual inspector biases, and gathering of more data. The disadvantages of large inspection teams from the laboratory's perspective are their intrusive nature on operations and their cost. The cost to the laboratory may vary from $1,800 to $12,500 per inspection (see Figure 8-3). Cost is also a consideration for employers who wish to inspect their laboratory. Depending on the number of inspectors and duration of the inspection, private inspections may cost employers several thousand dollars.

Laboratories take inspections seriously because their operations are

TABLE 8-8 Distribution of NIDA and CAP/AACC Checklist Questions and Critical Checklist Questions[15,16,21]

Checklist Section	% of Checklist Questions		Number of Critical Questions	
	NIDA	CAP/AACC	NIDA	CAP/AACC
1. Laboratory vitae	5%	5%	1	6
2. Testing procedures	3%	N/A[a]	2	N/A
3. SOP	13%	7%	3	0
4. Specimen handling	4%	5%	9	7
5. Records	7%	6%	9	4
6. Personnel	11%	3%	20	1
7. Reagents	2%	3%	0	0
8. Quality control	14%	3%	5	4
9. Reporting	6%	4%	15	1
10. Equipment & maintenance	6%	9%	0	0
11. Immunoassay	15%	5%	8	0
12. GC/MS	14%	5%	7	0
13. Quality assurance	N/A	4%	N/A	2
14. Performance testing	N/A	1%	N/A	1
15. TLC, HPLC, & GC	N/A	6%	N/A	1
16. Water quality	N/A	6%	N/A	0
17. Facilities	N/A	7%	N/A	0
18. Laboratory safety	N/A	24%	N/A	0
Totals	100%	99%	79	27

[a] Sections labeled N/A generally are included as portions of other checklist sections.

critically evaluated, and because they are costly. Inspections, by their nature, may place laboratories on the defensive. The qualifications and personalities of the inspectors are important in achieving an objective and thorough inspection. Inspection areas such as personnel qualifications require interviews, are subjective, and may be perceived as adversarial. An adversarial atmosphere is counterproductive and may threaten the professional relationship that must exist between the laboratory and its clients; therefore, discretion should be used in selecting inspectors and in performing inspections.

Resources

The inspection team must understand the range of services offered by the laboratory to determine if the laboratory has sufficient resources. This as-

Current NIDA fees to laboratories are as follows:

Applicant Laboratory		*Certified Laboratory*
Application fee	$ 750	
First PT	$ 920	PT – 6/year @ $460 = $2,760
Second PT	$ 920	
Third PT	$ 920	
Initial inspection	$12,500	Inspections – 2/year
3-month inspection	$12,500	@ $12,500 = $25,000
Total =	$28,510	Total = $27,760

Current CAP/AACC fees to laboratories are as follows:

Applicant Laboratory		*Certified Laboratory*
Application fee	$ 515[a]	
First PT	$ 188	PT –4/year @ $188 = $752
Second PT	$ 188	
Third PT	$ 188	
Initial inspection		Annual inspection
(1–500) specimens/day	$1,800	$1,800
(501–1000) specimens/day	$2,100	$2,100
(1001–2000) specimens/day	$2,400	$2,400
(>2000) specimens/day	$3,000	$3,000
Total =	[b]	Total = [b]

[a]Should the laboratory proceed with the PT and Inspection, the Application fee is credited toward PT and Inspection costs.
[b]Certification cost varies with the testing volume of the laboratory.

FIGURE 8-3. Fees for laboratory certification.

sessment includes human resources, scientific/technical expertise, facilities, and equipment.[20] There must be an adequate number of trained personnel to perform forensic urine drug testing. Expertise in immunoassay and GC/MS testing, chain of custody procedures, interpretation of data, and providing expert witness testimony is required. Facilities include the physical plant, security, and safety. Facilities, security, and equipment requirements are different in full-service laboratories than in those focused on forensic urine drug testing laboratories. If the laboratory performs workplace drug testing and clinical toxicology analyses, those staff assigned to clinical testing should not have access to the forensic areas. The physical plant must provide a safe working environment with adequate ventilation, safe storage facilities for hazardous chemicals, and secure facilities for handling biological samples. Security is required for storage of both analytical records and specimens.[9]

Standard Operating Procedures

Inspections begin with inspectors meticulously reading the laboratory's SOP manual. This manual must describe all of the laboratory's policies and procedures in detail. The SOP is the standard against which the laboratory's analytical data are judged during legal challenges. Inspectors score the SOP manual on thoroughness and on scientific and technical merit.

Specimen Handling

Inspectors review specimen handling protocols to ensure their adequacy. They evaluate specimen receiving procedures to see that only properly sealed and identified specimens are accepted for testing. The external chain of custody document—the form completed at the collection site—must reflect each person who handled the specimen or aliquot. The internal chain of custody document—the form the laboratory uses to track specimens after they are received—should contain the signature of each person handling the specimen and the date and reason for each handling. Each specimen aliquot used for analysis should have a similar chain of custody. To be acceptable, specimen handling and chain of custody documents must demonstrate that the specimens/aliquots were in the custody of an authorized person or stored in a secure area at all times.

Immunoassays

Immunoassay is the most widely accepted screening technology for workplace drug testing. Inspection of immunoassay is based on observing screening as it is being performed by the laboratory, interviewing analysts, and reviewing records. Inspectors observe testing to ensure that the proper reagents, calibrators, and controls are used. They also interview the analysts and certifying scientists to determine if they fully understand the theory and operational nuances of immunoassay screening. Inspectors review calibration, control, and PT records and determine if the screening analyses are being performed optimally. They also audit QA records of routine and nonroutine instrument maintenance data, QC and QC corrective actions, and reagent preparation and validation log books.

GC/MS

The inspectors' assessment of GC/MS competency is analogous to that used for immunoassay. They observe GC/MS testing while in progress, interview analysts, and review confirmation records. GC/MS is a more scientifically sophisticated analytic technique than immunoassay and, therefore, requires highly trained analysts. Technologies have improved GC/MS efficiency, and may allow technicians with only a superficial understanding of GC/MS to perform testing. This condition necessitates thorough interviews of analysts, certifying scientists, and the laboratory director to assess the depth of their GC/MS understanding. The NIDA Guidelines specify that GC/MS confirmation of positive screening findings be performed. They do not provide detailed GC/MS methods; therefore, inspectors review the laboratory's methods development records. These data should document the accuracy, reliability, and specificity of each method. Inspectors audit QA maintenance records for frequency of GC/MS calibration and cleaning, nonroutine repairs, and corrective actions for problems.

Reagents, Equipment, and Maintenance

Reagents, equipment, and equipment maintenance are inspected both independently and as part of the immunoassay and GC/MS areas. Inspectors check reagents to see that they are properly prepared, stored, verified, and that they are not used for testing past their expiration date. Inspectors examine the maintenance and calibration records for analytical balances and pipets as well as records of temperature-dependent reagent refrigerators, specimen storage refrigerators and freezers, and incubators.

QC

NIDA places an emphasis on the inspection of QC; as shown in Table 8-8, 14 percent of the checklist questions are about QC, and four of these are critical questions. Inspectors document that drug reference materials are purchased from reputable suppliers; that calibrators and controls are properly prepared, validated, and labeled; and that the laboratory adheres to the QC acceptance criteria defined in its SOP. Inspectors evaluate PT sample results in conjunction with internal QC records to identify testing trends that indicate potential analytical problems.

Personnel

Inspectors must thoroughly evaluate the laboratory's personnel. Inspectors should interview key personnel such as the director, certifying scientists, technical and quality control supervisors, and analysts. The quality and reliability of the laboratory's results are dependent on the staff's competency. Personnel deficiencies can be difficult to correct. If a key staff member is unqualified for his or her position, it may take months for the laboratory to recruit and train a suitable replacement. This is particularly true of the director because qualified directors are in short supply. The laboratory director is responsible for training of personnel, SOP writing, QA and QC procedures, methods development and validation, problem solving, and reporting.

Records

The records audit and the personnel interviews are the most revealing aspects of the inspection. Records reveal if the laboratory follows its SOP manual and if personnel judiciously complete their assignments. Records demonstrate if COC procedures are used effectively, if QC acceptance criteria are followed, if immunoassay and GC/MS analytical methods are accurate and reliable, and if the laboratory's reports are supported by in-house documents. Computerization, facsimile, and telephone data transmission make it essential to inspect the laboratory's reporting procedures. Reports must clearly identify the employee tested, drugs tested, date of specimen receipt and reporting, results, laboratory reporting official, and they must be transmitted confidentially. Unfortunately, inspectors usually have no means of ensuring that electronically transmitted reports are received in a secure area or that they undergo medical review.

THE FUTURE OF MONITORING LABORATORY PERFORMANCE

Many diverse factors affect drug testing and make predicting needs in monitoring laboratory performance difficult. The methods of monitoring will probably change only insofar as required to accommodate any changes in federal agency regulations, legislation, and sports medicine testing. The demand to monitor laboratory performance will follow the demand for testing.

NIDA and DOT have solicited recommendations to improve their regulations. On-site drug test analysis remains a controversial topic. It is prohibited by the current NIDA Guidelines, but permitted by the NRC regulations and by proposed federal law S. 1903.[28] In on-site drug testing, screening tests are performed at the collection site; presumptive positive specimens are sent elsewhere for GC/MS confirmation. The current NIDA requirements for GC/MS expertise, QC, GC/MS, QA, data review, record keeping and reporting would need major modification to accommodate on-site testing. Criteria for monitoring on-site testing facilities would need to be developed.

The Omnibus Transportation Employee Testing Act of 1991 requires DOT to design alcohol testing protocols by October 28, 1992.[29] Future legislation will also impact the future of drug testing and monitoring laboratory performance. Federal bills H.R. 33 and S. 1903 are being considered.[28,30] Both proposed laws would extend certain aspects of the NIDA Guidelines into private-sector workplace testing. S. 1903 permits on-site screening and expands the drug classes tested to include any controlled substances (Schedule I through V), alcohol, anabolic steroids, and illegally used prescription drugs. H.R. 33 also permits testing of additional drugs to include benzodiazepines, anabolic steroids, and Schedule I and II barbiturates. The inclusion of additional drug classes increases the complexity of testing and of monitoring laboratory performance. For example, NIDA-certified laboratories currently test for seven drugs in five drug classes. Sixty-three benzodiazepines have been synthesized and theoretically two thousand could be prepared.[31,32] Benzodiazepines are commonly prescribed,[33] are therapeutic at low doses, and appear in the urine at low concentrations. Because they are commonly prescribed, benzodiazepine tests would have relatively high positive rates. These characteristics require the laboratory to improve selectivity for identification, enhance sensitivity for detection, and have additional expertise to determine the parent drug source of the urinary metabolites. The demand to monitor laboratory performance will increase in scope and volume to satisfy the anticipated detection rates and complexity of testing for these additional drug classes.

The number of International Olympic Committee (IOC)-approved sports medicine testing laboratories has increased from 5 in 1980 to 22 in 1988.[34] CAP/AACC predicts that laboratories will find it difficult to keep pace with the growing demand for sports testing.[34] In total, over 100 compounds are banned by the IOC including anabolic steroids, drugs of abuse, narcotics, and diuretics.[34] This large number of drugs and drug classes places new demands on laboratories, MROs, and laboratory monitoring services. Laboratories have to design QC and QA programs to cover testing of additional drugs, external blind sample and PT challenges are needed to test the laboratory's proficiency at testing new drugs, and inspections need to incorporate additional oversight. Those involved in monitoring laboratories need to expand their expertise to accommodate the additional QC, QA, PT, external blind testing, and inspection needs of sports medicine testing.

A number of additional factors will interact to shape the future of workplace drug testing and consequently will impact laboratory monitoring. Increasing numbers of employers may begin drug testing programs, new state and local laws continue to modify the standards for testing, court rulings and arbitration awards impact testing programs, and the "war on drugs" will serve as a catalyst as long as it enjoys political support. The current legal and political support of workplace drug testing leads the authors to cautiously predict a continued increase in the demand for monitoring laboratory performance. The demand will be for an increased volume of services and an increase in the breadth of the services. Laboratories may be asked to test for a variety of new drug classes such as ethanol, anabolic steroids, benzodiazepines, barbiturates, and diuretics. This will expand the scope of laboratory services and monitoring requirements into many prescription drugs, into fitness-for-duty determinations, and into sports medicine.

SUMMARY

This chapter has described QC and QA (internal performance monitoring methods) and PT, external blind testing, and inspections (external performance monitoring methods). Together the internal and external performance monitoring programs constitute a thorough system of laboratory review. No single monitoring technique provides comprehensive laboratory oversight: Each has strengths and weaknesses.

1. Quality control ensures that analytical systems are performing within predefined limits as testing is being performed. Laboratory variations in establishing QC programs and QC acceptance criteria for analytical data preclude reliance on QC as the sole indicator of testing reliability.
2. Quality assurance programs provide internal oversight of all laboratory processes. The laboratory can use QA to diagnose, predict, and avoid operational problems. Unfortunately, many laboratories either misunderstand the function of QA or do not maintain sufficient records to utilize QA for its intended role.
3. Open performance testing results show the laboratory's analytical capabilities under ideal circumstances and provide important interlaboratory comparison data. However, open PT results are not indicative of the day-to-day accuracy and reliability of the laboratory.
4. Laboratory inspections provide the most comprehensive performance monitoring tool. Inspections suffer from a lack of uniformity in form, frequency, size of inspection teams, and inspector diligence.
5. External blind performance testing is the most accurate reflection of the laboratory's day-to-day testing accuracy. External blind testing should instill confidence in the employer and the donor in the accuracy of testing; nevertheless, the lack of regulation of vendors, mechanisms to capture data, and communication of constructive feedback leaves this potential unfulfilled.

Acknowledgments
The authors thank Merritt Birky, Ph.D., Thomas Jennison, M.S., Fran Urry, Ph.D., and Walt Vogl, Ph.D., for their assistance in editing this manuscript.

References
1. Dinovo, E. C., F. L. Gottschalk, H. B. McGuire, and J. F. Heiser. Forensic toxicology proficiency monitoring: Results, experiences, and comments. *J. Anal. Toxicol.* 1977;1:126-129.
2. Davis, K. H., R. L. Hawks, and R. V. Blanke. Assessment of laboratory quality in urine drug testing. *JAMA* 1988;260:1,749-1,754.
3. Hansen, H. J., S. P. Samuel, and J. Boone. Crisis in drug testing. *JAMA* 1985;253:2382-2387.
4. Frings, C. S., R. M. White, and D. J. Battaglia. Status of drugs-of-abuse testing in urine: An AACC study. *Clin. Chem.* 1987;33:1,683-1,686.
5. Crouch, D. J., D. E. Rollins, T. A. Jennison, and D. E. Moody. Criteria for the evaluation of occupational drug screening laboratories. In *Proceedings of the 24th International Meeting of the International Association of Forensic Toxicologists.* Banff, Canada; 1987:228-235.
6. Standefer, J. 1990. Results show problems near cutoff concentrations. *Forensic Urine Drug Testing* March 1991:7-8.
7. Shoemaker, M. J., M. Klein, and L. Sidemand. Drug abuse proficiency testing in Pennsylvania, 1972-1976. *J. Anal. Toxicol.* 1977;1:130-138.
8. Mason, M. F. Some realities and results of proficiency testing of laboratories performing toxicological analyses. *J. Anal. Toxicol.* 1981;5:201-208.
9. Department of Health and Human Services (DHHS). Mandatory guidelines for federal workplace drug testing programs. *Fed. Reg.* 1988;53(April 11):11,970-11,989.
10. College of American Pathologists. *Standards for Accreditation. Forensic Urine Drug Testing Laboratories.* Northfield, Ill.: 1990.
11. Caplan, Y. H. Monitoring laboratory performance. NIDA National Conference on Drug Abuse Research & Practice. Washington, D.C. Jan. 12-15, 1991.
12. Finkle, B. S., R. V. Blanke, and J. M. Walsh, (eds.). *Technical, Scientific and Procedural Issues of Employee Drug Testing.* Rockville, Md.: National Institute on Drug Abuse; 1990.
13. Crouch, D. J., and D. G. Wilkins. Managing the transition from traditional toxicology laboratories to certified laboratories: Operational needs. *Medical Laboratory Observer,* September 1991:58-60.
14. Blanke, R. V. Accuracy in Urinalysis. In R. L. Hawks and N. C. Chiang (eds.), *Urine Testing for Drugs of Abuse.* Rockville, Md.: National Institute on Drug Abuse, 1986:43-53.
15. College of American Pathologists. *Forensic Urine Drug Testing (Inspection Checklist).* Northfield, Ill. 1990.
16. U.S. Department of Transportation. Procedures for transportation workplace drug testing programs. *Fed. Reg.* 1989;54(Dec. 1):49,854-49,884.

17. Nuclear Regulatory Commission. Fitness-for-duty programs. *Fed. Reg.* 1989; 54(June 7):24,468–24,508.
18. Department of Health and Human Services (DHHS). Clinical laboratory improvement amendments of 1988. *Fed. Reg.* 1992;57(Feb. 28):7,002–7,243.
19. National Laboratory Certification Program. *Laboratory Inspection Report.* Research Triangle Park, N.C.: Research Triangle Institute; 1991.
20. National Institute on Drug Abuse. *Training (Manual) of the Laboratory Inspectors for the National Laboratory Certification Program.* Rockville, Md.: National Institute on Drug Abuse; 1991.
21. Shewart, W. A. *Economic Control of Quality Control of the Manufactured Product.* New York: Van Nostrand; 1931.
22. Westgard, J. O., and P. L. Barry. *Cost-Effective Quality Control and Quality Productivity of Analytical Processes.* Washington, D.C.: American Association for Clinical Chemistry; 1986.
23. Smith, M. L. Personal communication. Armed Forces Institute of Pathology, Washington D.C. 20306-6000. June 1991.
24. American Association for Clinical Chemistry. *An Employer's Guide to Evaluating Laboratory Performance.* Washington, D.C. 1989.
25. 21 CFR § 1308.24
26. 21 CFR § 862.
27. Nuclear Regulatory Commission. Fitness-for-duty programs. *Audit Checklist for 10 CFR Parts 2 and 26.* Nuclear Regulatory Commission, 11555 Rockville Pike, Rockville, Md. 1989.
28. Quality Assurance in the Private Sector Drug Testing Act of 1989. S. 1903, 101st Congress, 1st sess., sec. 1, 1989.
29. Omnibus Transportation Employee Testing Act of 1991. Pub law 102-143, 102nd Congress, October 28, 1991.
30. Drug Testing Quality Act. H.R. 33, 102nd Congress, 1st sess., sec. 1, 1991.
31. Schultz, H. *Benzodiazepines, a Handbook: Basic Data, Analytical Methods, Pharmacokinetics and Comprehensive Literature.* Heidelberg: Springer-Verlag; 1982; 1–439.
32. Gilman, A. G., T. W. Rall, A. S. Nies, and P. Taylor (eds.). *The Pharmacological Basis of Therapeutics. Eighth Edition.* Elmsford, N.Y.: Pergamon; 1990.
33. Simonsen, L. L. Top 200 drug of 1990. *Pharmacy Times.* April 1991;57–71.
34. Sample, B. Accrediting plans for sports testing. *Forensic Urine Drug Testing.* March 1991;1–5.

9
Case Studies

Robert Swotinsky, M.D., M.P.H.

These case studies in drug testing derive from actual situations. They are grouped according to issues that they address, as follows: programmatic issues, urine collection procedures, pharmacologic issues, the Medical Review Officer (MRO) function, risk management, and proficiency testing.

Actual dispositions follow the presentations of the cases. The reader should consider how he or she would handle each case before reading its outcome. The dispositions of these case studies are not intended to substitute for the individualized consideration that should be given to the handling and resolving of any specific issue in workplace drug testing.

PROGRAMMATIC ISSUES

Cases

1. An employer's drug testing procedures are consistent with the NIDA (National Institute on Drug Abuse) Guidelines. Specimens are analyzed for the NIDA-5 drugs and cutoffs. One specimen inadvertently is analyzed for a 10-drug panel that includes the NIDA-5 drugs and cutoffs. The laboratory is NIDA-certified and uses NIDA procedures for the NIDA-5 drugs. All the results are negative. Is the test valid? Would the test be valid if any results were laboratory-positive?

2. A medical examiner evaluating a truck driver's fitness as per DOT's Federal Highway Administration (FHWA) regulations wants the result of the

driver's FHWA-mandated drug test before making a determination. Is this reasonable?

3. An applicant's drug test result for Company X is verified positive. He is not offered the job. He is then, coincidentally, assigned to Company X by a temporary agency. Because of the positive drug test, Company X tells the temporary agency this worker is unacceptable. Should Company X tell the temporary agency why?

4. A donor's drug test is laboratory-positive for opiates. The MRO wants to examine the donor for clinical signs of opiate abuse, but the donor refuses. How should the MRO or employer handle this?

Dispositions

1. If the results are laboratory-negative, it would seem reasonable to consider the test valid, since invalidating it would serve little purpose. If any result was positive, the test should be reported as invalid because the error may make the test indefensible. The employer should be alerted, in any case, to the error, preferably without identifying the donor.

2. The medical examiner can request the result of the FHWA-mandated drug test, if performed, from the MRO. (The MRO would need the driver's authorization to release the results to the examiner.) The medical examiner should not withhold the medical determination if no result is provided. The medical exam and the DOT-mandated drug test can be separate and distinct activities. Drivers using particular drugs are medically unqualified under FHWA's medical standards. The physician can test a driver for drugs as part of the medical examination using any collection, or analytic protocol, at any cutoff levels [55 FR 3551, Feb. 1, 1990]. However, such testing can serve as a FHWA-mandated drug test only if it complies with FHWA drug testing regulations.

During a period before February 1990, the FHWA medical examiner's certificate asked the physician, with one signature, to certify that the driver was medically qualified *and* had passed the drug test. If the medical examiner did not have the drug test result, the reference to the drug test could be lined out. In February 1990, FHWA removed the drug test reference from the certificate and modified the medical examination form to allow the examiner to indicate if the results of a (DOT or non-DOT) drug test were considered as part of the medical examination. The employer, not the physician, is responsible for determining that the driver has passed both the medical exam and the Department of Transportation (DOT) drug test.

3. No. The donor's test result cannot be released to third parties without the donor's consent, except where required by regulations or by a court of law. It is highly unlikely the donor would consent to releasing information about a positive result.

4. The MRO cannot insist that the donor submit to an examination. The employer, however, can insist on this if the employer's policy states that, as a condition of employment, the donor must cooperate with the MRO's review.

URINE COLLECTION PROCEDURES
Cases

1. The laboratory receives a chain-of-custody block on a DOT form in which the collector's signature is missing from line 2 under "Released By" (as in Figure 9-1). Is the test valid?

2. Copy 3 of a DOT chain of custody form, the copy with the donor's name, is mistakenly sent to the laboratory. The laboratory notifies the MRO that it has canceled the test. The MRO believes the test is valid, and directs the laboratory to analyze the specimen. The laboratory result is positive, but the laboratory certifying scientist refuses to sign the certification statement on the chain of custody form. Is the test invalid because the donor's name has been revealed to the laboratory? Is the test valid without the certifying scientist's signature?

3. A donor has a verified positive DOT-mandated drug test and starts treatment for cocaine addiction. Subsequent scrutiny reveals that the donor's initials were missing from the bottle seal. Should the test be nullified?

4. A donor has kidney failure, is on dialysis, and produces no urine. How is this handled under the federal procedures?

5. A specimen bottle leaks in transit to the laboratory. Is the test valid?

	PURPOSE OF CHANGE	RELEASED BY	RECEIVED BY	DATE
1.	Provide Specimen for Test	- DONOR -	Jane Smith / janesmith	4/2/92
2.	Ship Specimen		Overnight Express	4/2/92
3.				

FIGURE 9-1. Urine collection procedures: Case 1.

Dispositions

1. The test is invalid. The missing collector's signature raises the possibility that someone else may have handled the specimen before shipment to the laboratory. Incomplete chain of custody blocks and other errors and omissions have met mixed outcomes in legal challenges. There is a growing consensus on which errors are significant enough to cause invalidation. DOT's recommended test invalidation criteria (presented in Chapter 5, Figure 5-4) provide guidance, but do not constitute regulation.

2. The accidental disclosure of a donor's name to the laboratory should not invalidate the test. The laboratory is unlikely to treat the specimen differently based on the name. This error does, however, breach the donor's confidentiality. A drug testing program that by design discloses the donors' names to the laboratory is in violation of the DOT Procedures.

A laboratory-positive result is valid in DOT testing only if the laboratory certifying scientist has signed the certification statement on the chain of custody form. In this case, the laboratory certifying scientist, after speaking with a DOT official, agreed to sign the certification statement and send it to the MRO.

3. Opinions sought from several sources were divided. This test's outcome, after much deliberation, was left unchanged.

4. Under the DOT model of drug testing, if a donor cannot produce a 60 ml specimen within an 8-hour period, the MRO refers the donor for an evaluation to determine if there is a medical basis for this. Upon completion of this evaluation, the MRO reports the conclusion to the employer. The examinee who produces no urine cannot undergo a urine test.

5. No. Material leaking out of the bottle raises the possibility that material leaked into the bottle and contaminated the specimen.

PHARMACOLOGIC ISSUES

Cases

1. A donor tests positive for morphine at 350 ng/ml. The donor took a legitimately prescribed medication that contained codeine, but codeine was not detected at the 300-ng/ml cutoff. Could the donor's medication have caused this result?

2. A truck driver has an accident and is treated for lacerations. His post-accident drug test is positive for benzoylecgonine. The driver denies having used cocaine. What potential legitimate source(s) of cocaine should be considered?

3. A law enforcement officer tests positive for methamphetamine. Seven days before the collection, the officer had participated in a seizure of a clandestine drug laboratory. During that seizure he handled methamphetamine-containing materials while gathering evidence. Could this exposure account for the result?

4. A donor tests positive for phenobarbital (a barbiturate). Prior to the test, he took the following over-the-counter medications: aspirin, Primatene P®, Tylenol®, Alka Seltzer®, and Nyquil®. Do any of these contain phenobarbital?

Dispositions

1. Yes. Codeine metabolizes to morphine. The test's outcome in this case was "negative." The most relevant additional information is the absolute codeine concentration—in this case, it was 280 ng/ml. The morphine-to-codeine ratio can help in exploring potential legitimate explanations for the results.[1] Nevertheless, the morphine-to-codeine ratio is not an infallible way to distinguish codeine from morphine use.[2]

2. TAC [*T*etracaine, *A*drenaline, and *C*ocaine] is a topical anesthetic typically used for repair of lacerations. Its use can cause cocaine-positive urine test results.[3]

3. No. Methamphetamine can be absorbed through mucosal surfaces and, to a lesser extent, through skin. Nevertheless, a single dose of methamphetamine typically produces positive urine tests for no more than 24 hours.[4] The officer's passive exposure seven days before the collection does not account for the positive test result.

Dermal absorption can also result in urinary concentrations of cocaine metabolite, but these concentrations have reportedly been below 300 ng/ml.[5,6]

4. Primatene P® tablets sold over the counter contain phenobarbital.

Some employers test for drugs beyond the NIDA-5, for example, for barbiturates, benzodiazepines, methaqualone. Because some of these drugs have common legitimate uses—both prescription and over the counter—testing for them results in a higher frequency of "laboratory-positive/MRO-negative" results.

THE MEDICAL REVIEW OFFICER FUNCTION

Cases

1. A donor's random DOT-mandated drug test is positive for codeine at 350 ng/ml. He tells the MRO he took his son's Robitussin A-C® for a cough

that interfered with his sleep the night before the test. Is this explanation legitimate?

2. A donor's specimen is laboratory-negative but dilute (specific gravity 1.001, creatinine concentration 0.16 g/L). At the donor's subsequent test, the specimen is collected under direct observation, and is also laboratory-negative and dilute (specific gravity 1.002, creatinine concentration 0.16 g/L). Is the first test valid? Is the second, witnessed test valid?

3. A urine specimen measures 80°F at the collection site. The micturition technician (collector), in accordance with the employer's policy, collects a second specimen under direct observation: It measures 97°F. Both specimens are submitted to the laboratory. The 80°F specimen is laboratory-positive for marijuana; the donor has no legitimate explanation. The 97°F specimen is laboratory-negative. How should these tests be reported to the employer?

4. A specimen is split between two laboratories. Lab A reports it as positive for 9-THC. Upon request, Lab B analyzes the specimen; Lab B detects 9-THC below the cutoff value. The donor has no legitimate explanation for marijuana use. How should this be reported to the employer? If Lab B were to find no 9-THC in the specimen, how should the test be reported to the employer?

5. An MRO receives and reviews chain-of-custody forms before reporting laboratory-negative DOT-mandated drug test results. Is this necessary? Is it necessary in nonregulated testing?

6. A donor's test is laboratory-positive for codeine and small amounts of 6-monoacetylmorphine. The donor says he ate two poppy seed bagels on the morning of the test. There are no needle tracks or other clinical signs of opiate abuse. How should the MRO report this result?

7. A donor's test is laboratory-positive for marijuana metabolite. The donor insists he never used marijuana and provides a statement from his wife supporting this claim. He angrily insists there must have been a mistake, and demands that a copy of the results be sent to his attorney. The attorney subsequently calls and asks for a copy of the records. Should the MRO report this result as negative, positive, or invalid?

8. A donor's test is laboratory-positive for marijuana metabolite. The donor insists he never used marijuana. He says he hunts and eats deer that graze where marijuana grows wild. Is this a plausible explanation?

9. A donor tests positive for benzodiazepines (oxazepam at 310 ng/ml, nordiazepam at 150 ng/ml). The donor had obtained a nonprescription health product by mail order, which he used for neck pain. The product, "NAN-LIEN CHUIFONG TOUKUWAN," is analyzed and found to contain diazepam. How should this test be reported to the employer?

10. The MRO block on the DOT-type chain of custody form reads:

> I have reviewed the laboratory results for the specimen identified by this form in accordance with applicable federal requirements. My final determination/verification is:
>
> *(Check one)*   ☐ *NEGATIVE*   ☐ *POSITIVE*
>
> SIGNATURE OF MEDICAL REVIEW OFFICER: _____ DATE: _____

Do the DOT Procedures require that the MRO complete this block?

Dispositions

1. No, per DOT's guidance to MROs: "An employee who acknowledges taking another individual's controlled substance prescribed medication has admitted unauthorized use of a controlled substance and the positive result should be verified positive."[1]

2. Both tests are valid. A dilute specimen, in DOT- or NRC-mandated testing, is a basis for performing a witnessed collection at the next test. However, dilute specimens are not a basis for invalidating the test.

3. The laboratory-negative test's outcome is "negative" and the laboratory-positive test's outcome is "verified positive." These are two individual tests, and their determinations are made independently. (In this instance, the MRO obtained quantitative 9-THC levels of each specimen. The laboratory-positive specimen had a level of 25 ng/ml. The laboratory-negative specimen had a level of 10 ng/ml.)

4. The test's outcome is "verified positive." The analysis of a split specimen need only demonstrate the presence of the drug or metabolite—without regard to the cutoff—to confirm the first specimen's laboratory-positive result. If the analysis of the split specimen found no 9-THC, the outcome of the test would be negative.

5. In DOT-mandated testing, the MRO or the MRO's designate must review the COC forms of negatives prior to reporting, and the MRO must review the COC forms of positives prior to reporting. Current MRO practices with respect to "review" of negatives vary widely, particularly in nonregulated testing. MRO review of negative results can provide a significant

quality control check, but can also result in reporting delays and increased labor, and consequently increased costs to the employer.

6. Positive. 6-Monoacetylmorphine is specific for heroin use.

7. Positive. This donor's behavior serves as a reminder that the workplace drug testing process can be fraught with emotion. The denial, anger, and legal threat do not alter the MRO's basic responsibility, which is to determine whether there is a legitimate medical explanation for the result.

8. No. (This was a real case!)

9. In the published case report,[7] the test was reported as "negative," the donor was advised to stop taking the medicine, and the employer was advised to retest the donor in 4 weeks. (Positive benzodiazepine drug tests have also been attributed to Nervo Tiamin Tranquilizante Vitaminado, which was obtained over the counter in Mexico,[8] and to Tung Shueh Pill, which was obtained over the counter in the United States.[9])

10. The DOT Procedures do not require completion of the MRO block on the chain of custody form. Many MROs complete this block on COC forms corresponding to laboratory-positive results. Those who do not should record their final determinations elsewhere in each drug test's file.

RISK MANAGEMENT

Cases

1. A donor's test is verified positive. Several months later the MRO receives from the donor's attorney a written request for the records of that test. How should the MRO respond?

2. An applicant tests positive for cocaine. While reviewing the result, the MRO learns that the donor is a police officer. The MRO determines the result is "verified positive" and reports it as such to the prospective employer. Should the MRO also notify the police officer's current employer? If the officer's current employer inquires as to the result of the test, how should the MRO respond?

3. An MRO is named in a suit alleging violation of the donor's Fourth Amendment's protection against unreasonable searches. The MRO's professional liability insurance carrier denies coverage for this claim, and states that the physician's policy does not cover constitutional torts. How could this have been avoided?

4. An MRO's staff gathers data and makes preliminary decisions regarding each laboratory-positive drug test result. A staff person then informs the donor of this decision. If the decision is "verified-positive," the staff person advises the donor that he/she has the option of speaking with the MRO, but suggests that this is unlikely to change the outcome. Is this practice acceptable?

Dispositions

1. The MRO should first notify the employer. In federally mandated testing, the employer must give donors access to their drug test records upon written request. The employer's policy may also address this. If the employer chooses to release the records, the MRO may send copies to the employer, who in turn can send them to the donor. This routing helps the employer control the records, and it may protect the MRO from potential allegations of mishandling the request.

2. The results may be released to the police department only if the donor authorizes this, or if required by law. Each donor should complete a written authorization, prior to the test, that releases the results to the employer and the MRO. In the unusual event that an applicant has agreed to release the results to his or her current employer, that information should be communicated by the employer, not by the MRO. The MRO responded to inquiries from the police department in this case by agreeing to discuss the information if subpoenaed to do so.

3. The MRO should have determined beforehand whether the policy covered drug testing services. If not, the MRO might have supplemented the existing policy with a rider that provides supplemental coverage. Or, the physician could have provided MRO services through a separate corporation, which could have been insured with general liability insurance.

4. No. The federal procedures allow the MRO's staff to participate in reviews. Each donor must have an opportunity to discuss his/her result directly with the MRO. The donor–MRO interview is at the core of the MRO function, and must not be circumvented by suggesting to the donor that it is unnecessary.

PERFORMANCE TESTING
Cases

1. A performance testing specimen containing 1,500 ng/ml of methamphetamine tests laboratory-negative. The laboratory's toxicologist says the

methamphetamine level was ~1500 ng/ml, but the laboratory reported the result as negative because no amphetamine was detected. Why does the laboratory make methamphetamine-positive results contingent on the co-presence of amphetamine?

2. External blind performance testing specimens are prepared with analytes at concentrations 5 percent above the screening cutoff levels. Over time, most of these specimens screen negative. What can be concluded about accuracy?

Dispositions

1. NIDA requires that, in order to report a methamphetamine-positive result (i.e., a methamphetamine level above 500 ng/ml), the specimen must also contain amphetamine, a metabolite of methamphetamine, at a level equal to or greater than 200 ng/ml. Methamphetamine use produces both methamphetamine and amphetamine in the urine. This requirement was enacted in 1990 when NIDA learned that ephedrine could, under certain conditions, cause drug test results that were false positive for methamphetamine but negative for amphetamine.[10]

2. This data does not allow a conclusion about the laboratory's accuracy. The external blind performance testing specimen concentrations are too low. They should each be at least 20 percent above the cutoff level of the assay (screening or confirmatory) that is being monitored. Some analytes may have lower than expected concentrations because they degrade over time or adsorb to the sides of the specimen container. Also, some degree of analytic variability is to be expected in the quantitative values that are obtained.

References
1. U.S. Department of Transportation. *Medical Review Officer Guide*. Washington, D.C. 1990.
2. Cone, E. J., P. Welch, B. D. Paul, and J. M. Mitchell. Forensic drug testing for opiates, III. Urinary excretion rates of morphine and codeine following codeine administration. *J. Anal. Toxicol.* 1991;15:161–166.
3. Schwartz, R. H., M. Altieri, and S. Bogema. Topical anesthesia using TAC (tetracaine, adrenalin, and cocaine) produces "dirty urine." *Otolaryngol. Head Neck Surg.* 1990;102:200–201.
4. Hawks, R. L., and C. N. Chiang. Examples of specific drug assays. In R. L. Hawks and C. N. Chiang (eds.), *Urine Testing for Drugs of Abuse*. Rockville, Md.: National Institute on Drug Abuse, 1986:95.

5. Baselt, R. C., J. Y. Chang, and D. M. Yoshikawa. On the dermal absorption of cocaine. *J. Anal. Toxicol.* 1990;14:383–384.
6. El Sohly, M. A. Urinalysis and casual handling of marijuana and cocaine. *J. Anal. Toxicol.* 1991;15:46.
7. Dupont, R. L., and S. C. Bogema. Benzodiazepines in a health catalog product. *JAMA* 1990;264:695.
8. Fedoruk, M. J., and L. Lee. Positive preemployment urine drug screen caused by foreign-manufactured vitamin formulation. *West. J. Med.,* in press.
9. Floren, A. E., and W. Fitter. Contamination of urine with diazepam and mefanamic acid from an oriental remedy. *J. Occup. Med.* 1991;33:1,168–1,169.
10. Autry, J. A. *Notice to All DHHS/NIDA Certified Laboratories.* Rockville, Md.: National Institute on Drug Abuse; December 19, 1990.

Appendix A

Drug Screening in the Workplace: Ethical Guidelines

The following features should be included in any program for the screening of employees and prospective employees for drugs:

1. A written company policy and procedure concerning drug use and screening for the presence of drugs should exist and be applied impartially.

2. The reason for any requirement for screening for drugs should be clearly documented. Such reasons might involve safety for the individual, other employees, or the public; security needs; or requirements related to job performance.

3. Affected employees and applicants should be informed in advance about the company's policy concerning drug use and screening. They should be made aware of their right to refuse such screening and the consequences of such refusal to their employment.

4. Where special safety or security needs justify testing for drugs on an announced and possibly random basis, employees should be made aware in advance that this will be done from time to time. Care should be taken to assure that such tests are done in a uniform and impartial manner for all employees in the affected group(s).

5. Written consent for screening and for communication of results to the employer should be obtained from each individual prior to screening.

Source: Adapted from American College of Occupational Medicine. Drug screening in the workplace: Ethical Guidelines. *J. Occup. Med.* 1991;33:651–652.

6. Collection, transportation, and analysis of the specimens and the reporting of the results should meet stringent legal, technical, and ethical requirements. The process should be under the supervision of a licensed physician.

7. A licensed physician who is qualified as a Medical Review Officer (MRO) should evaluate positive results prior to a report being made to the employer. This may require the obtaining of supplemental information from the employee or applicant in order to ensure that a positive test does not represent appropriate use of prescription drugs, over-the-counter medication, or other substances that could cause a positive test.

Training of the MRO should include the pharmacology of substance abuse, laboratory testing methodology and quality control, forensic toxicology, pertinent federal regulations, legal and ethical requirements, chemical dependency illness, employee assistance programs, and rehabilitation.

8. The affected employee or applicant should be advised of positive results by the physician and have the opportunity for explanation and discussion prior to the reporting of results to the employer, if feasible. The mechanism for accomplishing this should be clearly defined.

9. The employee or applicant having indication of a drug abuse problem should be advised concerning appropriate treatment resources.

10. Any report to the employer should provide only the information needed for work placement purposes or as required by government regulations. Identification to the employer of the particular drug(s) found and quantitative levels is not necessary, unless required by law. Reports to the employer should be made by a physician sensitive to the various considerations involved.

Appendix B

Mandatory Guidelines for Federal Workplace Drug Testing Programs

MANDATORY GUIDELINES FOR FEDERAL WORKPLACE DRUG TESTING PROGRAMS

Subpart A—General
1.1 Applicability.
1.2 Definitions.
1.3 Future Revisions.

Subpart B—Scientific and Technical Requirements
2.1 The Drugs.
2.2 Specimen Collection Procedures.
2.3 Laboratory Personnel.
2.4 Laboratory Analysis Procedures.
2.5 Quality Assurance and Quality Control.
2.6 Interim Certification Procedures.
2.7 Reporting and Review of Results.
2.8 Protection of Employee Records.
2.9 Individual Access to Test and Laboratory Certification Results.

Subpart C—Certification of Laboratories Engaged in Urine Drug Testing for Federal Agencies
3.1 Introduction.
3.2 Goals and Objectives of Certification.
3.3 General Certification Requirements.
3.4 Capability to Test for Five Classes of Drugs.
3.5 Initial and Confirmatory Capability at Same Site.
3.6 Personnel.
3.7 Quality Assurance and Quality Control.
3.8 Security and Chain of Custody.
3.9 One-Year Storage for Confirmed Positives.
3.10 Documentation.
3.11 Reports.
3.12 Certification.
3.13 Revocation.
3.14 Suspension.
3.15 Notice; Opportunity for Review.
3.16 Recertification.
3.17 Performance Test Requirement for Certification.
3.18 Performance Test Specimen Composition.
3.19 Evaluation of Performance Testing.
3.20 Inspections.
3.21 Results of Inadequate Performance.

Authority: E.O. 12564 and sec. 503 of Pub. L. 100-71.

Subpart A—General

1.1 Applicability.

(a) These mandatory guidelines apply to:

(1) Executive Agencies as defined in 5 U.S.C. 105;
(2) The Uniformed Services, as defined in 5 U.S.C. 2101 (3) (but excluding the Armed Forces as defined in 5 U.S.C. 2101(2));
(3) And any other employing unit or authority of the Federal Government except the United States Postal Service, the Postal Rate Commission, and employing units or authorities in the Judicial and Legislative Branches.

(b) Any agency or component of an agency with a drug testing program in existence as of September 15, 1986, and the Departments of Transportation and Energy shall take such action as may be necessary to ensure that the agency is brought into compliance with these Guidelines no later than 90 days after they take effect, except that any judicial challenge that affects these Guidelines shall not affect drug testing programs subject to this paragraph.

(c) Except as provided in 2.6, Subpart C of these Guidelines (which establishes laboratory certification standards) applies to any laboratory which has or seeks certification to perform urine drug testing for Federal agencies under a drug testing program conducted under E.O. 12564. Only laboratories certified under these standards are authorized to perform urine drug testing for Federal agencies.

(d) The Intelligence Community, as defined by Executive Order No. 12333, shall be subject to these Guidelines only to the extent agreed to by the head of the affected agency.

(e) These Guidelines do not apply to drug testing conducted under legal authority other than E.O. 12564, including testing of persons in the criminal justice system, such as arrestees, detainees, probationers, incarcerated persons, or parolees.

(f) Agencies may not deviate from the provisions of these Guidelines without the written approval of the Secretary. In requesting approval for a deviation, an agency must petition the Secretary in writing and describe the specific provision or provisions for which a deviation is sought and the rationale therefor. The Secretary may approve the request upon a finding of good cause as determined by the Secretary.

1.2 Definitions.

For purposes of these Guidelines the following definitions are adopted:

Aliquot A portion of a specimen used for testing.

Chain of Custody Procedures to account for the integrity of each urine specimen by tracking its handling and storage from point of specimen collection to final disposition of the specimen. These procedures shall require that an approved agency chain of custody form be used from time of collection to receipt by the laboratory and that upon receipt of the laboratory an appropriate laboratory chain of custody form(s) account for the sample or sample aliquots within the laboratory. Chain of custody forms shall, at a minimum, include an entry documenting date and purpose each time a specimen or aliquot is handled or transferred and identifying every individual in the chain of custody.

Collection Site A place designated by the agency where individuals present themselves for the purpose of providing a specimen of their urine to be analyzed for the presence of drugs.

Collection Site Person A person who instructs and assists individuals at a collection site and who receives and makes an initial examination of the urine specimen provided by those individuals. A collection site person shall have successfully completed training to carry out this function.

Confirmatory Test A second analytical procedure to identify the presence of a specific drug or metabolite which is independent of the initial test and which uses a different technique and chemical principle from that of the initial test in order to ensure reliability and accuracy. (At this time gas chromatography/mass spectrometry (GC/MS) is the only authorized confirmation method for cocaine, marijuana, opiates, amphetamines, and phencyclidine.)

Initial Test (also known as Screening Test) An immunoassay screen to eliminate "negative" urine specimens from further consideration.

Medical Review Officer A licensed physician responsible for receiving laboratory results generated by an agency's drug testing program who has knowledge of substance abuse disorders and has appropriate medical training to interpret and evaluate an individual's positive test result together with his or her medical history and any other relevant biomedical information.

Permanent Record Book A permanently bound book in which identifying data on each specimen collected at a collection site are permanently recorded in the sequence of collection.

Reason to Believe Reason to believe that a particular individual may alter or substitute the urine specimen as provided in section 4(c) of E.O. 12564.

Secretary The Secretary of Health and Human Services or the Secretary's designee. The Secretary's designee may be contractor or other recognized organization which acts on behalf of the Secretary in implementing these Guidelines.

1.3 Future Revisions.

In order to ensure the full reliability and accuracy of drug assays, the accurate reporting of test results, and the integrity and efficacy of Federal drug testing programs, the Secretary may make changes to these Guidelines to reflect improvements in the available science and technology. These changes will be published in final as a notice in the **Federal Register**.

Subpart B—Scientific and Technical Requirements

2.1 The Drugs.

(a) The President's Executive Order 12564 defines "illegal drugs" as those included in Schedule I or II of the Controlled Substances Act (CSA), but not when used pursuant to a valid prescription or when used as otherwise authorized by law. Hundreds of drugs are covered under Schedule I and II and while it is not feasible to test routinely for all of them. Federal drug testing programs shall test for drugs as follows:

(1) Federal agency applicant and random drug testing programs shall at a minimum test for marijuana and cocaine;

(2) Federal agency applicant and random drug testing programs are also authorized to test for opiates, amphetamines, and phencyclidine; and

(3) When conducting reasonable suspicion, accident, or unsafe practice testing, a Federal agency may test for any drug listed in Schedule I or II of the CSA.

(b) Any agency covered by these guidelines shall petition the Secretary in writing for approval to include in its testing protocols any drugs (or classes of drugs) not listed for Federal agency testing in paragraph (a) of this section. Such approval shall be limited to the use of the appropriate science and technology and shall not otherwise limit agency discretion to test for any drugs covered under Schedule I or II of the CSA.

(c) Urine specimens collected pursuant to Executive Order 12564, Pub. L. 100-71, and these Guidelines shall be used only to test for those drugs included in agency drug-free workplace plans and may not be used to conduct any other analysis or test unless otherwise authorized by law.

(d) These Guidelines are not intended to limit any agency which is specifically authorized by law to include additional categories of drugs in the drug testing of its own employees or employees in its regulated industries.

2.2 Specimen Collection Procedures.

(a) Designation of Collection Site. Each agency drug testing program shall have one or more designated collection sites which have all necessary personnel, materials, equipment, facilities, and supervision to provide for the collection, security, temporary storage, and shipping or transportation of urine specimens to a certified drug testing laboratory.

(b) Security Procedures shall provide for the designated collection site to be secure. If a collection site facility is dedicated solely to urine collection, it shall be secure at all times. If a facility cannot be dedicated solely to drug testing, the portion of the facility used for testing shall be secured during drug testing.

(c) Chain of Custody. Chain of custody standardized forms shall be properly executed by authorized collection site personnel upon receipt of specimens. Handling and transportation of urine specimens from one authorized individual or place to another shall always be accomplished through chain of custody procedures. Every effort shall be made to minimize the number of persons handling specimens.

(d) Access to Authorized Personnel Only. No unauthorized personnel shall be permitted in any part of the designated collection site when urine specimens are collected or stored.

(e) Privacy. Procedures for collecting urine specimens shall allow individual privacy unless there is reason to believe that a particular individual may alter or substitute the specimen to be provided.

(f) Integrity and Identity of Specimen. Agencies shall take precautions to ensure that a urine specimen not be adulterated or diluted during the collection procedure and that information on the urine bottle and in the record book can identify the individual from whom the specimen was collected. The following minimum precautions shall be taken to ensure that unadulterated specimens are obtained and correctly identified:

(1) To deter the dilution of specimens at the collection site, toilet bluing agents shall be placed in toilet tanks wherever possible, so the reservoir of water in the toilet bowl always remains blue. There shall be no other source of water (e.g., no shower or sink) in the enclosure where urination occurs.

(2) When an individual arrives at the collection site, the collection site person shall request the individual to present photo identification. If the individual does not have proper photo identification, the collection site person shall contact the supervisor of the individual, the coordinator of the drug testing program, or any other agency official who can positively identify the individual. If the individual's identity cannot be established, the collection site person shall not proceed with the collection.

(3) If the individual fails to arrive at the assigned time, the collection site person shall contact the appropriate authority to obtain guidance on the action to be taken.

(4) The collection site person shall ask the individual to remove any unnecessary outer garments such as a coat or jacket that might conceal items or substances that could be used to tamper with or adulterate the individual's urine specimen. The collection site person shall ensure that all personal belongings such as a purse or briefcase remain with the outer garments. The individual may retain his or her wallet.

(5) The individual shall be instructed to wash and dry his or her hands prior to urination.

(6) After washing hands, the individual shall remain in the presence of the collection site person and shall not have access to any water fountain, faucet, soap dispenser, cleaning agent or

any other materials which could be used to adulterate the specimen.

(7) The individual may provide his/her specimen in the privacy of a stall or otherwise partitioned area that allows for individual privacy.

(8) The collection site person shall note any unusual behavior or appearance in the permanent record book.

(9) In the exceptional event that an agency-designated collection site is not accessible and there is an immediate requirement for specimen collection (e.g., an accident investigation), a public rest room may be used according to the following procedures: A collection site person of the same gender as the individual shall accompany the individual into the public rest room which shall be made secure during the collection procedure. If possible, a toilet bluing agent shall be placed in the bowl and any accessible toilet tank. The collection site person shall remain in the rest room, but outside the stall, until the specimen is collected. If no bluing agent is available to deter specimen dilution, the collection site person shall instruct the individual not to flush the toilet until the specimen is delivered to the collection site person. After the collection site person has possession of the specimen, the individual will be instructed to flush the toilet and to participate with the collection site person in completing the chain of custody procedures.

(10) Upon receiving the specimen from the individual, the collection site person shall determine that it contains at least 60 milliliters of urine. If there is less than 60 milliliters of urine in the container, additional urine shall be collected in a separate container to reach a total of 60 milliliters. (The temperature of the partial specimen in each separate container shall be measured in accordance with paragraph (f)(12) of this section, and the partial specimens shall be combined in one container.) The individual may be given a reasonable amount of liquid to drink for this purpose (e.g., a glass of water). If the individual fails for any reason to provide 60 milliliters of urine, the collection site person shall contact the appropriate authority to obtain guidance on the action to be taken.

(11) After the specimen has been provided and submitted to the collection site person, the individual shall be allowed to wash his or her hands.

(12) Immediately after the specimen is collected, the collection site person shall measure the temperature of the specimen. The temperature measuring device used must accurately reflect the temperature of the specimen and not contaminate the specimen. The time from urination to temperature measurement is critical and in no case shall exceed 4 minutes.

(13) If the temperature of a specimen is outside the range of 32.5°–37.7°C/90.5°–99.8°F, that is a reason to believe that the individual may have altered or substituted the specimen, and another specimen shall be collected under direct observation of a same gender collection site person and both specimens shall be forwarded to the laboratory for testing. An individual may volunteer to have his or her oral temperature taken to provide evidence to counter the reason to believe the individual may have altered or substituted the specimen caused by the specimen's temperature falling outside the prescribed range.

(14) Immediately after the specimen is collected, the collection site person shall also inspect the specimen to determine its color and look for any signs of contaminants. Any unusual findings shall be noted in the permanent record book.

(15) All specimens suspected of being adulterated shall be forwarded to the laboratory for testing.

(16) Whenever there is reason to believe that a particular individual may alter or substitute the specimen to be provided, a second specimen shall be obtained as soon as possible under the direct observation of a same gender collection site person.

(17) Both the individual being tested and the collection site person shall keep the specimen in view at all times prior to its being sealed and labeled. If the specimen is transferred to a second bottle, the collection site person shall request the individual to observe the transfer of the specimen and the placement of the tamperproof seal over the bottle cap and down the sides of the bottle.

(18) The collection site person and the individual shall be present at the same time during procedures outlined in paragraphs (f)((19)–(f)(22) of this section.

(19) The collection site person shall place securely on the bottle an identification label which contains the date, the individual's specimen number, and any other identifying information provided or required by the agency.

(20) The individual shall initial the identification label on the specimen bottle for the purpose of certifying that it is the specimen collected from him or her.

(21) The collection site person shall enter in the permanent record book all information identifying the specimen. The collection site person shall sign the permanent record book next to the identifying information.

(22) The individual shall be asked to read and sign a statement in the permanent record book certifying that the specimen identified as having been collected from him or her is in fact that specimen he or she provided.

(23) A higher level supervisor shall review and concur in advance with any decision by a collection site person to obtain a specimen under the direct observation of a same gender collection site person based on a reason to believe that the individual may alter or substitute the specimen to be provided.

(24) The collection site person shall complete the chain of custody form.

(25) The urine specimen and chain of custody form are now ready for shipment. If the specimen is not immediately prepared for shipment, it shall be appropriately safeguarded during temporary storage.

(26) While any part of the above chain of custody procedures is being performed, it is essential that the urine specimen and custody documents be under the control of the involved collection site person. If the involved collection site person leaves his or her work station momentarily, the specimen and custody form shall be taken with him or her or shall be secured. After the collection site person returns to the work station, the custody process will continue. If the collection site person is leaving for an extended period of time, the specimen shall be packaged for mailing before he or she leaves the site.

(g) Collection Control. To the maximum extent possible, collection site personnel shall keep the individual's specimen bottle within sight both before and after the individual has urinated. After the specimen is collected, it shall be properly sealed and labeled. An approved chain of custody form shall be used for maintaining control and accountability of each specimen from the point of collection to final disposition of the specimen. The date and purpose shall be documented on an approved chain of custody form each time a specimen is handled or transferred and every individual in the chain shall be identified. Every effort shall be made to minimize the number of persons handling specimens.

(h) Transportation to Laboratory. Collection site personnel shall arrange to ship the collected specimens to the drug testing laboratory. The specimens shall be placed in containers designed to minimize the possibility of damage during shipment, for example, specimen boxes or padded mailers; and those containers shall be securely sealed to eliminate the possibility of undetected tampering. On the tape sealing the

container, the collection site supervisor shall sign and enter the date specimens were sealed in the containers for shipment. The collection site personnel shall ensure that the chain of custody documentation is attached to each container sealed for shipment to the drug testing laboratory.

2.3 *Laboratory Personnel.*

(a) Day-to-Day Management.

(1) The laboratory shall have a qualified individual to assume professional, organizational, educational, and administrative responsibility for the laboratory's urine drug testing facility.

(2) This individual shall have documented scientific qualifications in analytical forensic toxicology. Minimum qualifications are:

(i) Certification as a laboratory director by the State in forensic or clinical laboratory toxicology; or

(ii) A Ph.D. in one of the natural sciences with an adequate undergraduate and graduate education in biology, chemistry, and pharmacology or toxicology, or

(iii) Training and experience comparable to a Ph.D. in one of the natural sciences, such as a medical or scientific degree with additional training and laboratory/research experience in biology, chemistry, and pharmacology or toxicology; and

(iv) In addition to the requirements in (i), (ii), and (iii) above, minimum qualifications also require:

(A) Appropriate experience in analytical forensic toxicology including experience with the analysis of biological material for drugs of abuse, and

(B) Appropriate training and/or experience in forensic applications of analytical toxicology, e.g., publications, court testimony, research concerning analytical toxicology of drugs of abuse, or other factors which qualify the individual as an expert witness in forensic toxicology.

(3) This individual shall be engaged in and responsible for the day-to-day management of the drug testing laboratory even where another individual has overall responsibility for an entire multispecialty laboratory.

(4) This individual shall be responsible for ensuring that there are enough personnel with adequate training and experience to supervise and conduct the work of the drug testing laboratory. He or she shall assure the continued competency of laboratory personnel by documenting their inservice training, reviewing their work performance, and verifying their skills.

(5) This individual shall be responsible for the laboratory's having a procedure manual which is complete, up-to-date, available for personnel performing tests, and followed by those personnel. The procedure manual shall be reviewed, signed, and dated by this responsible individual whenever procedures are first placed into use or changed or when a new individual assumes responsibility for management of the drug testing laboratory. Copies of all procedures and dates on which they are in effect shall be maintained. (Specific contents of the procedure manual are described in 2.4(n)(1).)

(6) This individual shall be responsible for maintaining a quality assurance program to assure the proper performance and reporting of all test results; for maintaining acceptable analytical performance for all controls and standards; for maintaining quality control testing; and for assuring and documenting the validity, reliability, accuracy, precision, and performance characteristics of each test and test system.

(7) This individual shall be responsible for taking all remedial actions necessary to maintain satisfactory operation and performance of the laboratory in response to quality control systems not being within performance specifications, errors in result reporting or in analysis of performance testing results. This individual shall ensure that sample results are not reported until all corrective actions have been taken and he or she can assure that the tests results provided are accurate and reliable.

(b) Test Validation. The laboratory's urine drug testing facility shall have a qualified individual(s) who reviews all pertinent data and quality control results in order to attest to the validity of the laboratory's test reports. A laboratory may designate more than one person to perform this function. This individual(s) may be any employee who is qualified to be responsible for day-to-day management or operation of the drug testing laboratory.

(c) Day-to-Day Operations and Supervision of Analysts. The laboratory's urine drug testing facility shall have an individual to be responsible for day-to-day operations and to supervise the technical analysts. This individual(s) shall have at least a bachelor's degree in the chemical or biological sciences or medical technology or equivalent. He or she shall have training and experience in the theory and practice of the procedures used in the laboratory, resulting in his or her thorough understanding of quality control practices and procedures; the review, interpretation, and reporting of test results; maintenance of chain of custody; and proper remedial actions to be taken in response to test systems being out of control limits or detecting aberrant test or quality control results.

(d) Other Personnel. Other technicians or nontechnical staff shall have the necessary training and skills for the tasks assigned.

(e) Training. The laboratory's urine drug testing program shall make available continuing education programs to meet the needs of laboratory personnel.

(f) Files. Laboratory personnel files shall include: resume of training and experience; certification or license, if any; references; job descriptions; records of performance evaluation and advancement; incident reports; and results of tests which establish employee competency for the position he or she holds, such as a test for color blindness, if appropriate.

2.4 *Laboratory Analysis Procedures.*

(a) Security and Chain of Custody. (1) Drug testing laboratories shall be secure at all times. They shall have in place sufficient security measures to control access to the premises and to ensure that no unauthorized personnel handle specimens or gain access to the laboratory processes or to areas where records are stored. Access to these secured areas shall be limited to specifically authorized individuals whose authorization is documented. With the exception of personnel authorized to conduct inspections on behalf of Federal agencies for which the laboratory is engaged in urine testing or on behalf of the Secretary, all authorized visitors and maintenance and service personnel shall be escorted at all times. Documentation of individuals accessing these areas, dates, and time of entry and purpose of entry must be maintained.

(2) Laboratories shall use chain of custody procedures to maintain control and accountability of specimens from receipt through completion of testing, reporting of results, during storage, and continuing until final disposition of specimens. The date and purpose shall be documented on an appropriate chain of custody form each time a specimen is handled or transferred, and every individual in the chain shall be identified. Accordingly, authorized technicians shall be responsible for each urine specimen or aliquot in their possession and shall sign and complete chain of custody forms for those specimens or aliquots as they are received.

(b) Receiving. (1) When a shipment of specimens is received, laboratory personnel shall inspect each package for evidence of possible tampering and compare information on specimen bottles within each package to the information on the accompanying chain of custody forms. Any direct evidence of tampering or discrepancies in the information on specimen bottles and the agency's chain of custody forms attached to the shipment shall be immediately reported to the agency and shall be noted on the laboratory's chain of custody form which shall accompany the specimens while they are in the laboratory's possession.

(2) Specimen bottles will normally be retained within the laboratory's accession area until all analyses have been completed. Aliquots and the laboratory's chain of custody forms shall be used by laboratory personnel for conducting initial and confirmatory tests.

(c) Short-Term Refrigerated Storage. Specimens that do not receive an initial test within 7 days of arrival at the laboratory shall be placed in secure refrigeration units. Temperatures shall not exceed 6°C. Emergency power equipment shall be available in case of prolonged power failure.

(d) Specimen Processing. Laboratory facilities for urine drug testing will normally process specimens by grouping them into batches. The number of specimens in each batch may vary significantly depending on the size of the laboratory and its workload. When conducting either initial or confirmatory tests, every batch shall contain an appropriate number of standards for calibrating the instrumentation and a minimum of 10 percent controls. Both quality control and blind performance test samples shall appear as ordinary samples to laboratory analysts.

(e) Initial Test. (1) The initial test shall use an immunoassay which meets the requirements of the Food and Drug Administration for commercial distribution. The following initial cutoff levels shall be used when screening specimens to determine whether they are negative for these five drugs or classes of drugs:

	Initial test level (ng/ml)
Marijuana metabolites	100
Cocaine metabolites	300
Opiate metabolites	¹ 300
Phencyclidine	25
Amphetamines	1,000

¹ 25ng/ml if immunoassay specific for free morphine.

(2) These test levels are subject to change by the Department of Health and Human Services as advances in technology or other considerations warrant identification of these substances at other concentrations. Initial test methods and testing levels for other drugs shall be submitted in writing by the agency for the written approval of the Secretary.

(f) Confirmatory Test. (1) All specimens identified as positive on the initial test shall be confirmed using gas chromatography/mass spectrometry (GC/MS) techniques at the cutoff values listed in this paragraph for each drug. All confirmations shall be by quantitative analysis. Concentrations which exceed the linear region of the standard curve shall be documented in the laboratory record as "greater than highest standard curve value."

	Confirmatory test level (ng/ml)
Marijuana metabolite ¹	15
Cocaine metabolite ²	150
Opiates:	
Morphine	* 300
Codeine	* 300
Phencyclidine	25
Amphetamines:	
Amphetamine	500
Methamphetamine	500

¹ Delta-9-tetrahydrocannabinol-9-carboxylic acid.
² Benzoylecgonine.

(2) These test levels are subject to change by the Department of Health and Human Services as advances in technology or other considerations warrant identification of these substances at other concentrations. Confirmatory test methods and testing levels for other drugs shall be submitted in writing by the agency for the written approval of the Secretary.

(g) Reporting Results. (1) The laboratory shall report test results to the agency's Medical Review Officer within an average of 5 working days after receipt of the specimen by the laboratory. Before any test result is reported (the results of initial tests, confirmatory tests, or quality control data), it shall be reviewed and the test certified as an accurate report by the responsible individual. The report shall identify the drugs/metabolites tested for, whether positive or negative, and the cutoff for each, the specimen number assigned by the agency, and the drug testing laboratory specimen identification number. The results (positive and negative) for all specimens submitted at the same time to the laboratory shall be reported back to the Medical Review Officer at the same time.

(2) The laboratory *shall report as negative* all specimens which are negative on the initial test or negative on the confirmatory test. Only specimens confirmed positive shall be reported positive for a specific drug.

(3) The Medical Review Officer may request from the laboratory and the laboratory shall provide quantitation of test results. The Medical Review Officer may not disclose quantitation of test results to the agency but shall report only whether the test was positive or negative.

(4) The laboratory may transmit results to the Medical Review Officer by various electronic means (for example, teleprinters, facsimile, or computer) in a manner designed to ensure confidentiality of the information. Results may not be provided verbally by telephone. The laboratory must ensure the security of the data transmission and limit access to any data transmission, storage, and retrieval system.

(5) The laboratory shall send only to the Medical Review Officer a certified copy of the original chain of custody form signed by the individual responsible for day-to-day management of the drug testing laboratory or the individual responsible for attesting to the validity of the test reports.

(6) The laboratory shall provide to the agency official responsible for coordination of the drug-free workplace program a monthly statistical summary of urinalysis testing of Federal employees and shall not include in the summary any personal identifying information. Initial and confirmation data shall be included from test results reported within that month. Normally this summary shall be forwarded by registered or certified mail not more than 14 calendar days after the end of the month covered by the summary. The summary shall contain the following information:

(i) Initial Testing:
(A) Number of specimens received;
(B) Number of specimens reported out; and
(C) Number of specimens screened positive for:
  Marijuana metabolites
  Cocaine metabolites
  Opiate metabolites
  Phencyclidine
  Amphetamines

(ii) Confirmatory Testing:
(A) Number of specimens received for confirmation;
(B) Number of specimens confirmed positive for:
  Marijuana metabolite

Cocaine metabolite
Morphine, codeine
Phencyclidine
Amphetamine
Methamphetamine

(7) The laboratory shall make available copies of all analytical results for Federal drug testing programs when requested by DHHS or any Federal agency for which the laboratory is performing drug testing services.

(8) Unless otherwise instructed by the agency in writing, all records pertaining to a given urine specimen shall be retained by the drug testing laboratory for a minimum of 2 years.

(h) Long-Term Storage. Long-term frozen storage ($-20\ °C$ or less) ensures that positive urine specimens will be available for any necessary retest during administrative or disciplinary proceedings. Unless otherwise authorized in writing by the agency, drug testing laboratories shall retain and place in properly secured long-term frozen storage for a minimum of 1 year all specimens confirmed positive. Within this 1-year period an agency may request the laboratory to retain the specimen for an additional period of time, but if no such request is received the laboratory may discard the specimen after the end of 1 year, except that the laboratory shall be required to maintain any specimens under legal challenge for an indefinite period.

(i) Retesting Specimens. Because some analytes deteriorate or are lost during freezing and/or storage, quantitation for a retest is not subject to a specific cutoff requirement but must provide data sufficient to confirm the presence of the drug or metabolite.

(j) Subcontracting. Drug testing laboratories shall not subcontract and shall perform all work with their own personnel and equipment unless otherwise authorized by the agency. The laboratory must be capable of performing testing for the five classes of drugs (marijuana, cocaine, opiates, phencyclidine, and amphetamines) using the initial immunoassay and confirmatory GC/MS methods specified in these Guidelines.

(k) Laboratory Facilities. (1) Laboratory facilities shall comply with applicable provisions of any State licensure requirements.

(2) Laboratories certified in accordance with Subpart C of these Guidelines shall have the capability, at the same laboratory premises, of performing initial and confirmatory tests for each drug or metabolite for which service is offered.

(l) Inspections. The Secretary, any Federal agency utilizing the laboratory, or any organization performing laboratory certification on behalf of the Secretary shall reserve the right to inspect the laboratory at any time. Agency contracts with laboratories for drug testing, as well as contracts for collection site services, shall permit the agency to conduct unannounced inspections. In addition, prior to the award of a contract the agency shall carry out preaward inspections and evaluation of the procedural aspects of the laboratory's drug testing operation.

(m) Documentation. The drug testing laboratories shall maintain and make available for at least 2 years documentation of all aspects of the testing process. This 2-year period may be extended upon written notification by DHHS or by any Federal agency for which laboratory services are being provided: The required documentation shall include personnel files on all individuals authorized to have access to specimens; chain of custody documents; quality assurance/quality control records; procedure manuals; all test data (including calibration curves and any calculations used in determining test results); reports; performance records on performance testing; performance on certification inspections; and hard copies of computer-generated data. The laboratory shall be required to maintain documents for any specimen under legal challenge for an indefinite period.

(n) Additional Requirements for Certified Laboratories.—*(1) Procedure Manual.* Each laboratory shall have a procedure manual which includes the principles of each test, preparation of reagents, standards and controls, calibration procedures, derivation of results, linearity of methods, sensitivity of the methods, cutoff values, mechanisms for reporting results, controls, criteria for unacceptable specimens and results, remedial actions to be taken when the test systems are outside of acceptable limits, reagents and expiration dates, and references. Copies of all procedures and dates on which they are in effect shall be maintained as part of the manual.

(2) Standards and Controls. Laboratory standards shall be prepared with pure drug standards which are properly labeled as to content and concentration. The standards shall be labeled with the following dates: when received; when prepared or opened; when placed in services; and expiration date.

(3) Instruments and Equipment. (i) Volumetric pipettes and measuring devices shall be certified for accuracy or be checked by gravimetric, colorimetric, or other verification procedure. Automatic pipettes and dilutors shall be checked for accuracy and reproducibility before being placed in service and checked periodically thereafter.

(ii) There shall be written procedures for instrument set-up and normal operation, a schedule for checking critical operating characteristics for all instruments, tolerance limits for acceptable function checks and instructions for major trouble shooting and repair. Records shall be available on preventive maintenance.

(4) Remedial Actions. There shall be written procedures for the actions to be taken when systems are out of accceptable limits or errors are detected. There shall be documentation that these procedures are followed and that all necessary corrective actions are taken. There shall also be in place systems to verify all stages of testing and reporting and documentation that these procedures are followed.

(5) Personnel Available To Testify at Proceedings. A laboratory shall have qualified personnel available to testify in an administrative or disciplinary proceeding against a Federal employee when that proceeding is based on positive urinalysis results reported by the laboratory.

2.5 *Quality Assurance and Quality Control.*

(a) General. Drug testing laboratories shall have a quality assurance program which encompasses all aspects of the testing process including but not limited to specimen acquisition, chain of custody, security and reporting of results, initial and confirmatory testing, and validation of analytical procedures. Quality assurance procedures shall be designed, implemented, and reviewed to monitor the conduct of each step of the process of testing for drugs.

(b) Laboratory Quality Control Requirements for Initial Tests. Each analytical run of specimens to be screened shall include:

(1) Urine specimens certified to contain no drug;

(2) Urine specimens fortified with known standards; and

(3) Positive controls with the drug or metabolite at or near the threshold (cutoff).

In addition, with each batch of samples a sufficient number of standards shall be included to ensure and document the linearity of the assay method over time in the concentration area of the cutoff. After acceptable values are obtained for the known standards, those values will be used to calculate sample data. Implementation of procedures to ensure that carryover does not contaminate the testing of an individual's specimen shall be documented. A minimum of 10 percent of all test samples shall be quality control specimens. Laboratory quality control samples, prepared from spiked urine samples of determined concentration shall be included in the run and should appear as normal samples to laboratory analysts. One percent of each run, with a minimum of at least one sample, shall be the laboratory's own quality control samples.

(c) Laboratory Quality Control Requirements for Confirmation Tests. Each analytical run of specimens to be confirmed shall include:

(1) Urine specimens certified to contain no drug;

(2) Urine specimens fortified with known standards; and

(3) Positive controls with the drug or metabolite at or near the threshold (cutoff).

The linearity and precision of the method shall be periodically documented. Implementation of procedures to ensure that carryover does not contaminate the testing of an individual's specimen shall also be documented.

(d) Agency Blind Performance Test Procedures. (1) Agencies shall purchase drug testing services only from laboratories certified by DHHS or a DHHS-Recognized certification program in accordance with these Guidelines. Laboratory participation is encouraged in other performance testing surveys by which the laboratory's performance is compared with peers and reference laboratories.

(2) During the initial 90-day period of any new drug testing program, each agency shall submit blind performance test specimens to each laboratory it contracts with in the amount of at least 50 percent of the total number of samples submitted (up to a maximum of 500 samples) and thereafter a minimum of 10 percent of all samples (to a maximum of 250) submitted per quarter.

(3) Approximately 80 percent of the blind performance test samples shall be blank (i.e., certified to contain no drug) and the remaining samples shall be positive for one or more drugs per sample in a distribution such that all the drugs to be tested are included in approximately equal frequencies of challenge. The positive samples shall be spiked only with those drugs for which the agency is testing.

(4) The Secretary shall investigate any unsatisfactory performance testing result and, based on this investigation, the laboratory shall take action to correct the cause of the unsatisfactory performance test result. A record shall be make of the Secretary's investigative findings and the corrective action taken by the laboratory, and that record shall be dated and signed by the individuals responsible for the day-to-day management and operation of the drug testing laboratory. Then the Secretary shall send the document to the agency contracting officer as a report of the unsatisfactory performance testing incident. The Secretary shall ensure notification of the finding to all other Federal agencies for which the laboratory is engaged in urine drug testing and coordinate any necessary action.

(5) Should a false positive error occur on a blind performance test specimen and the error is determined to be an administrative error (clerical, sample mixup, etc.), the Secretary shall require the laboratory to take corrective action to minimize the occurrence of the particular error in the future; and, if there is reason to believe the error could have been systematic, the Secretary may also require review and reanalysis of previously run specimens.

(6) Should a false positive error occur on a blind performance test specimen and the error is determined to be a technical or methodological error, the laboratory shall submit all quality control data from the batch of specimens which included the false positive specimen. In addition, the laboratory shall retest all specimens analyzed positive for that drug or metabolite from the time of final resolution of the error back to the time of the last satisfactory performance test cycle. This retesting shall be documented by a statement signed by the individual responsible for day-to-day management of the laboratory's urine drug testing. The Secretary may require an on-site review of the laboratory which may be conducted unannounced during any hours of operations of the laboratory. The Secretary has the option of revoking (3.13) or suspending (3.14) the laboratory's certification or recommending that no further action be taken if the case is one of less serious error in which corrective action has already been taken, thus reasonably assuring that the error will not occur again.

2.6 Interim Certification Procedures.

During the interim certification period as determined under paragraph (c), agencies shall ensure laboratory competence by one of the following methods:

(a) Agencies may use agency or contract laboratories that have been certified for urinalysis testing by the Department of Defense; or

(b) Agencies may develop interim self-certification procedures by establishing preaward inspections and performance testing plans approved by DHHS.

(c) The period during which these interim certification procedures will apply shall be determined by the Secretary. Upon noticed by the Secretary that these interim certification procedures are no longer available, all Federal agencies subject to these Guidelines shall only use laboratories that have been certified in accordance with Subpart C of these Guidelines and all laboratories approved for interim certification under paragraphs (a) and (b) of this section shall become certified in accordance with Subpart C within 120 days of the date of this notice.

2.7 Reporting and Review of Results.

(a) Medical Review Officer Shall Review Results. An essential part of the drug testing program is the final review of results. A positive test result does not automatically identify an employee/applicant as an illegal drug user. An individual with a detailed knowledge of possible alternate medical explanations is essential to the review of results. This review shall be performed by the Medical Review Officer prior to the transmission of results to agency administrative officials.

(b) Medical Review Officer— Qualifications and Responsibilities. The Medical Review Officer shall be a licensed physician with knowledge of substance abuse disorders and may be an agency or contract employee. The role of the Medical Review Officer is to review and interpret positive test results obtained through the agency's testing program. In carrying out this responsibility, the Medical Review Officer shall examine alternate medical explanations for any positive test result. This action could include conducting a medical interview with the individual, review of the individual's medical history, or review of any other relevant biomedical factors. The Medical Review Officer shall review all medical records made available by the tested individual when a confirmed positive test could have resulted from legally prescribed medication. The Medical Review Officer shall not, however, consider the results of urine samples that are not obtained or processed in accordance with these Guidelines.

(c) Positive Test Result. Prior to making a final decision to verify a positive test result, the Medical Review Officer shall give the individual an opportunity to discuss the test result

with him or her. Following verification of a positive test result, the Medical Review Officer shall refer the case to the agency Employee Assistance Program and to the management official empowered to recommend or take administrative action.

(d) Verification for opiates; review for prescription mediation. Before the Medical Review Officer verifies a confirmed positive result for opiates, he or she shall determine that there is clinical evidence—in addition to the urine test—of illegal use of any opium, opiate, or opium derivative (e.g., morphine/codeine) listed in Schedule I or II of the Controlled Substances Act. (This requirement does not apply if the agency's GC/MS confirmation testing for opiates confirms the presence of 6-monoacetylmorphine.)

(e) Reanalysis Authorized. Should any question arise as to the accuracy or validity of a positive test result, only the Medical Review Officer is authorized to order a reanalysis of the original sample and such retests are authorized only at laboratories certified under these Guidelines.

(f) Result Consistent with Legal Drug Use. If the Medical Review Officer determines there is a legitimate medical explanation for the positive test result, he or she shall determine that the result is consistent with legal drug use and take no further action.

(g) Result Scientifically Insufficient. Additionally, the Medical Review Officer, based on review of inspection reports, quality control data, multiple samples, and other pertinent results, may determine that the result is scientifically insufficient for further action and declare the test specimen negative. In this situation the Medical Review Officer may request reanalysis of the original sample before making this decision. (The Medical Review Officer may request that reanalysis be performed by the same laboratory or, as provided in 2.7(e), that an aliquot of the original specimen be sent for reanalysis to an alternate laboratory which is certified in accordance with these Guidelines.) The laboratory shall assist in this review process as requested by the Medical Review Officer by making available the individual responsible for day-to-day management of the urine drug testing laboratory or other employee who is a forensic toxicologist or who has equivalent forensic experience in urine drug testing, to provide specific consultation as required by the agency. The Medical Review Officer shall report to the Secretary all negative findings based on scientific insufficiency but shall not include any personal identifying information in such reports.

2.8 Protection of Employee Records.

Consistent with 5 U.S.C. 522a(m) and 48 CFR 24.101–24.104, all laboratory contracts shall require that the contractor comply with the Privacy Act, 5 U.S.C. 552a. In addition, laboratory contracts shall require compliance with the patient access and confidentiality provisions of section 503 of Pub. L. 100–71. The agency shall establish a Privacy Act System of Records or modify an existing system, or use any applicable Government-wide system of records to cover both the agency's and the laboratory's records of employee urinalysis results. The contract and the Privacy Act System shall specifically require that employee records be maintained and used with the highest regard for employee privacy.

2.9 Individual Access to Test and Laboratory Certification Results.

In accordance with section 503 of Pub. L. 100–71, any Federal employee who is the subject of a drug test shall, upon written request, have access to any records relating to his or her drug test and any records relating to the results of any relevant certification, review, or revocation-of-certification proceedings.

Subpart C—Certification of Laboratories Engaged in Urine Drug Testing for Federal Agencies

3.1 Introduction.

Urine drug testing is a critical component of efforts to combat drug abuse in our society. Many laboratories are familiar with good laboratory practices but may be unfamiliar with the special procedures required when drug test results are used in the employment context. Accordingly, the following are minimum standards to certify laboratories engaged in urine drug testing for Federal agencies. Certification, even at the highest level, does not guarantee accuracy of each result reported by a laboratory conducting urine drug testing for Federal agencies. Therefore, results from laboratories certified under these Guidelines must be interpreted with a complete understanding of the total collection, analysis, and reporting process before a final conclusion is made.

3.2 Goals and Objectives of Certification.

(a) Uses of Urine Drug Testing. Urine drug testing is an important tool to identify drug users in a variety of settings. In the proper context, urine drug testing can be used to deter drug abuse in general. To be a useful tool, the testing procedure must be capable of detecting drugs or their metabolites at concentrations indicated in 2.4 (e) and (f).

(b) Need to Set Standards; Inspections. Reliable discrimination between the presence, or absence, of specific drugs or their metabolites is critical, not only to achieve the goals of the testing program but to protect the rights of the Federal employees being tested. Thus, standards have been set which laboratories engaged in Federal employee urine drug testing must meet in order to achieve maximum accuracy of test results. These laboratories will be evaluated by the Secretary or the Secretary's designee as defined in 1.2 in accordance with these Guidelines. The qualifying evaluation will involve three rounds of performance testing plus on-site inspection. Maintenance of certification requires participation in an every-other-month performance testing program plus periodic, on-site inspections. One inspection following successful completion of a performance testing regimen is required for initial certification. This must be followed by a second inspection within 3 months, after which biannual inspections will be required to maintain certification.

(c) Urine Drug Testing Applies Analytical Forensic Toxicology. The possible impact of a positive test result on an individual's livelihood or rights, together with the possibility of a legal challenge of the result, sets this type of test apart from most clinical laboratory testing. In fact, urine drug testing should be considered a special application of analytical forensic toxicology. That is, in addition to the application of appropriate analytical methodology, the specimen must be treated as evidence, and all aspects of the testing procedure must be documented and available for possible court testimony. Laboratories engaged in urine drug testing for Federal agencies will require the services and advice of a qualified forensic toxicologist, or individual with equivalent qualifications (both training and experience) to address the specific needs of the Federal drug testing program, including the demands of chain of custody of specimens, security, property documentation of all records, storage of positive specimens for later or independent testing, presentation of evidence in court, and expert witness testimony.

3.3 General Certification Requirements.

A laboratory must meet all the pertinent provisions of these Guidelines in order to qualify for certification under these standards.

3.4 Capability to Test for Five Classes of Drugs.

To be certified, a laboratory must be capable of testing for at least the following five classes of drugs: Marijuana, cocaine, opiates, amphetamines, and phencyclidine, using the initial immunoassay and quantitative confirmatory GC/MS methods specified in these Guidelines. The certification program will be limited to the five classes of drugs (2.1(a) (1) and (2)) and the methods (2.4 (e) and (f)) specified in these Guidelines. The laboratory will be surveyed and performance tested only for these methods and drugs. Certification of a laboratory indicates that any test result reported by the laboratory for the Federal Government meets the standards in these Guidelines for the five classes of using the methods specified. Certified laboratories must clearly inform non-Federal clients when procedures followed for those clients conform to the standards specified in these Guidelines.

3.5 Initial and Confirmatory Capability at Same Site.

Certified laboratories shall have the capability, at the same laboratory site, of performing both initial immunoassays and confirmatory GC/MS tests (2.4 (e) and (f)) for marijuana, cocaine, opiates, amphetamines, and phencyclidine and for any other drug or metabolite for which agency drug testing is authorized (2.1(a) (1) and (2)). All positive initial test results shall be confirmed prior to reporting them.

3.6 Personnel.

Laboratory personnel shall meet the requirements specified in 2.3 of these Guidelines. These Guidelines establish the exclusive standards for qualifying or certifying those laboratory personnel involved in urinalysis testing whose functions are prescribed by these Guidelines. A certification of a laboratory under these Guidelines shall be a determination that these qualification requirements have been met.

3.7 Quality Assurance and Quality Control.

Drug testing laboratories shall have a quality assurance program which encompasses all aspects of the testing process, including but not limited to specimen acquisition, chain of custody, security and reporting of results, initial and confirmatory testing, and validation of analytical procedures. Quality control procedures shall be designed, implemented, and reviewed to monitor the conduct of each step of the process of testing for drugs as specified in 2.5 of these Guidelines.

3.8 Security and Chain of Custody.

Laboratories shall meet the security and chain of custody requirements provided in 2.4(a).

3.9 One-Year Storage for Confirmed Positives.

All confirmed positive specimens shall be retained in accordance with the provisions of 2.4(h) of these Guidelines.

3.10 Documentation.

The laboratory shall maintain and make available for at least 2 years documentation in accordance with the specifications in 2.4(m).

3.11 Reports.

The laboratory shall report test results in accordance with the specifications in 2.4(g).

3.12 Certification.

(a) General. The Secretary may certify any laboratory that meets the standards in these Guidelines to conduct urine drug testing. In addition, the Secretary may consider to be certified and laboratory that is certified by a DHHS-recognized certification program in accordance with these Guidelines.

(b) Criteria. In determining whether to certify a laboratory or to accept the certification of a DHHS-recognized certification program in accordance with these Guidelines, the Secretary shall consider the following criteria:

(1) The adequacy of the laboratory facilities;
(2) The expertise and experience of the laboratory personnel;
(3) The excellence of the laboratory's quality assurance/quality control program;
(4) The performance of the laboratory on any performance tests;
(5) The laboratory's compliance with standards as reflected in any laboratory inspections; and
(6) Any other factors affecting the reliability and accuracy of drug tests and reporting done by the laboratory.

3.13 Revocation.

(a) General. The Secretary shall revoke certification of any laboratory certified under these provisions or accept revocation by a DHHS-recognized certification program in accordance with these Guidelines if the Secretary determines that revocation is necessary to ensure the full reliability and accuracy of drug tests and the accurate reporting of test results.

(b) Factors to Consider. The Secretary shall consider the following factors in determining whether revocation is necessary:

(1) Unsatisfactory performance in analyzing and reporting the results of drug tests; for example, a false positive error in reporting the results of an employee's drug test;
(2) Unsatisfactory participation in performance evaluations or laboratory inspections;
(3) A material violation of a certification standard or a contract term or other condition imposed on the laboratory by a Federal agency using the laboratory's services;
(4) Conviction for any criminal offense committed as an incident to operation of the laboratory; or
(5) Any other cause which materially affects the ability of the laboratory to ensure the full reliability and accuracy of drug tests and the accurate reporting of results.

(c) Period and Terms. The period and terms of revocation shall be determined by the Secretary and shall depend upon the facts and circumstances of the revocation and the need to ensure accurate and reliable drug testing of Federal employees.

3.14 Suspension.

(a) Criteria. Whenever the Secretary has reason to believe that revocation may be required and that immediate action is necessary in order to protect the interests of the United States and its employees, the Secretary may immediately suspend a laboratory's certification to conduct urine drug testing for Federal agencies. The Secretary may also accept suspension of certification by a DHHS-recognized certification program in accordance with these Guidelines.

(b) Period and Terms. The period and terms of suspension shall be determined by the Secretary and shall depend upon the facts and circumstances of the suspension and the need to ensure accurate and reliable drug testing of Federal employees.

3.15 Notice; Opportunity for Review.

(a) Written Notice. When a laboratory is suspended or the Secretary seeks to revoke certification, the Secretary shall immediately serve the laboratory with written notice of the suspension or proposed revocation by personal service or registered or certified mail, return

receipt requested. This notice shall state the following:
(1) The reasons for the suspension or proposed revocation;
(2) The terms of the suspension or proposed revocation; and
(3) The period of suspension or proposed revocation.

(b) Opportunity for Informal Review. The written notice shall state that the laboratory will be afforded an opportunity for an informal review of the suspension or proposed revocation if it so requests in writing within 30 days of the date of mailing or service of the notice. The review shall be by a person or persons designated by the Secretary and shall be based on written submissions by the laboratory and the Department of Health and Human Services and, at the Secretary's discretion, may include an opportunity for an oral presentation. Formal rules of evidence and procedures applicable to proceedings in a court of law shall not apply. The decision of the reviewing official shall be final.

(c) Effective Date. A suspension shall be effective immediately. A proposed revocation shall be effective 30 days after written notice is given or, if review is requested, upon the reviewing official's decision to uphold the proposed revocation. If the reviewing official decides not to uphold the suspension or proposed revocation, the suspension shall terminate immediately and any proposed revocation shall not take effect.

(d) DHHS-Recognized Certification Program. The Secretary's responsibility under this section may be carried out by a DHHS-recognized certification program in accordance with these Guidelines.

3.16 Recertification.

Following the termination or expiration of any suspension or revocation, a laboratory may apply for recertification. Upon the submission of evidence satisfactory to the Secretary that the laboratory is in compliance with these Guidelines or any DHHS-recognized certification program in accordance with these Guidelines, and any other conditions imposed as part of the suspension or revocation, the Secretary may recertify the laboratory or accept the recertification of the laboratory by a DHHS-recognized certification program.

3.17 Performance Test Requirement for Certification.

(a) An Initial and Continuing Requirement. The performance testing program is a part of the initial evaluation of a laboratory seeking certification (both performance testing and laboratory inspection are required) and of the continuing assessment of laboratory performance necessary to maintain this certification.

(b) Three Initial Cycles Required. Successful participation in three cycles of testing shall be required before a laboratory is eligible to be considered for inspection and certification. These initial three cycles (and any required for recertification) can be compressed into a 3-month period (one per month).

(c) Six Challenges Per Year. After certification, laboratories shall be challenged every other month with one set of at least 10 specimens a total of six cycles per year.

(d) Laboratory Procedures Identical for Performance Test and Routine Employee Specimens. All procedures associated with the handling and testing of the performance test specimens by the laboratory shall to the greatest extent possible be carried out in a manner identical to that applied to routine laboratory specimens, unless otherwise specified.

(e) Blind Performance Test. Any certified laboratory shall be subject to blind performance testing (see 2.5(d)). Performance on blind test specimens shall be at the same level as for the open or non-blind performance testing.

(f) Reporting—Open Performance Test. The laboratory shall report results of open performance tests to the certifying organization in the same manner as specified in 2.4(g)(2) for routine laboratory specimens.

3.18 Performance Test Specimen Composition.

(a) Description of the Drugs. Performance test specimens shall contain those drugs and metabolites which each certified laboratory must be prepared to assay in concentration ranges that allow detection of the analyte by commonly used immunoassay screening techniques. These levels are generally in the range of concentrations which might be expected in the urine of recent drug users. For some drug analytes, the specimen composition will consist of the parent drug as well as major metabolites. In some cases, more than one drug class may be included in one specimen container, but generally no more than two drugs will be present in any one specimen in order to imitate the type of specimen which a laboratory normally encounters. For any particular performance testing cycle, the actual composition of kits going to different laboratories will vary but, within any annual period, all laboratories participating will have analyzed the same total set of specimens.

(b) Concentrations. Performance test specimens shall be spiked with the drug classes and their metabolites which are required for certifications: marijuana, cocaine, opiates, amphetamines, and phencyclidine, with concentration levels set at least 20 percent above the cutoff limit for either the initial assay or the confirmatory test, depending on which is to be evaluated. Some performance test specimens may be identified for GC/MS assay only. Blanks shall contain less than 2 ng/ml of any of the target drugs. These concentration and drug types may be changed periodically in response to factors such as changes in detection technology and patterns of drug use.

3.19 Evaluation of Peformance Testing.

(a) Initial Certification. (1) An applicant laboratory shall not report any false positive result during performance testing for initial certification. Any false positive will automatically disqualify a laboratory from further consideration.

(2) An applicant laboratory shall maintain an overall grade level of 90 percent for the three cycles of performance testing required for initial certification, i.e., it must correctly identify and confirm 90 percent of the total drug challenges for each shipment. Any laboratory which achieves a score on any one cycle of the initial certification such that it can no longer achieve a total grade of 90 percent over the three cycles will be immediately disqualified from further consideration.

(3) An applicant laboratory shall obtain quantitative values for at least 80 percent of the total drug challenges which are ±20 percent or ±2 standard deviations of the calculated reference group mean (whichever is larger). Failure to achieve 80 percent will result in disqualification.

(4) An applicant laboratory shall not obtain any quantitative values that differ by more than 50 percent from the calculated reference group mean. Any quantitative values that differ by more than 50 percent will result in disqualification.

(5) For any individual drug, an applicant laboratory shall successfully detect and quantitate in accordance with paragraphs (a)(2), (a)(3), and (a)(4) of this section at least 50 percent of the total drug challenges. Failure to successfully quantitate at least 50 percent of the challenges for any individual drug will result in disqualification.

(b) Ongoing Testing of Certified Laboratories.—(1) False Positives and Procedures for Dealing With Them. No

false drug identifications are acceptable for any drugs for which a laboratory offers service. Under some circumstances a false positive test may result in suspension or revocation of certification. The most serious false positives are by drug class, such as reporting THC in a blank specimen or reporting cocaine in a specimen known to contain only opiates. Misidentifications within a class (e.g., codeine for morphine) are also false positives which are unacceptable in an appropriately controlled laboratory, but they are clearly less serious errors than misidentification of a class. The following procedures shall be followed when dealing with a false positive:

(i) The agency detecting a false positive error shall immediately notify the laboratory and the Secretary of any such error.

(ii) The laboratory shall provide the Secretary with a written explanation of the reasons for the error within 5 working days. If required by paragraph (b)(1)(v) below, this explanation shall include the submission of all quality control data from the batch of specimens that included the false positive specimen.

(iii) The Secretary shall review the laboratory's explanation within 5 working days and decide what further action, if any, to take.

(iv) If the error is determined to be an administrative error (clerical, sample mixup, etc.), the Secretary may direct the laboratory to take corrective action to minimize the occurence of the particular error in the future and, if there is reason to believe the error could have been systematic, may require the laboratory to review and reanalyze previously run specimens.

(v) If the error is determined to be technical or methodological error, the laboratory shall submit to the Secretary all quality control data from the batch of specimens which included the false positive specimen. In addition, the laboratory shall retest all specimens analyzed positive by the laboratory from the time fo final resolution of the error back to the time of the last satisfactory performance test cycle. This retesting shall be documented by a statement signed by the individual responsible for the day-to-day management of the laboratory's urine drug testing. Depending on the type of error which caused the false positive, this retesting may be limited to one analyte or may include any drugs a laboratory certified under these Guidelines must be prepared to assay. The laboratory shall immediately notify the agency if any result on a retest sample must be corrected because the critieria for a positive are not satisfied. The Secretary may suspend or revoke the laboratory's certification for all drugs or for only the drug or drug class in which the error occurred. However, if the case is one of a less serious error for which effective corrections have already been made, thus reasonably assuring that the error will not occur again, the Secretary may decide to take no further action.

(vi) During the time required to resolve the error, the laboratory shall remain certified but shall have a designation indicating that a false positive result is pending resolution. If the Secretary determines that the laboratory's certification must be suspended or revoked, the laboratory's official status will become "Suspended" or "Revoked" until the suspension or revocation is lifted or any recertification process is complete.

(2) *Requirement to Identify and Confirm 90 Percent of Total Drug Challenges.* In order to remain certified, laboratories must successfully complete six cycles of performance testing per year. Failure of a certified laboratory to maintain a grade of 90 percent on any required performance test cycle, i.e., to identify 90 percent of the total drug challenges and to correctly confirm 90 percent of the total drug challenges, may result in suspension or revocation of certification.

(3) *Requirement to Quantitate 80 Percent of Total Drug Challenges at ±20 Percent or ±2 standard deviations.* Quantitative values obtained by a certified laboratory for at least 80 percent of the total drug challenges must be ±20 percent or ±2 standard deviations of the calculated reference group mean (whichever is larger).

(4) *Requirement to Quantitate within 50 Percent of Calculated Reference Group Mean.* No quantitative values obtained by a certified laboratory may differ by more than 50 percent from the calculated reference group mean.

(5) *Requirement to Successfully Detect and Quantitate 50 Percent of the Total Drug Challenges for Any Individual Drug.* For any individual drug, a certified laboratory must successfully detect and quantitate in accordance with paragraphs (b)(2), (b)(3), and (b)(4) of this section at least 50 percent of the total drug challenges.

(6) *Procedures When Requirements in Paragraphs (b)(2)-(b)(5) of this Section Are Not Met.* If a certified laboratory fails to maintain a grade of 90 percent per test cycle after initial certification as required by paragraph (b)(2) of this section or if it fails to successfully quantitate results as required by paragraphs (b)(3), (b)(4), or (b)(5) of this section, the laboratory shall be immediately informed that its performance fell under the 90 percent level or that it failed to successfully quantitate test results and how it failed to successfully quantitate. The laboratory shall be allowed 5 working days in which to provide any explanation for its unsuccessful performance, including administrative error or methodological error, and evidence that the source of the poor performance has been corrected. The Secretary may revoke or suspend the laboratory's certification or take no further action, depending on the seriousness of the errors and whether there is evidence that the source of the poor performance has been corrected and that current performance meets the requirements for a certified laboratory under these Guidelines. The Secretary may require that additional performance tests be carried out to determine whether the source of the poor performance has been removed. If the Secretary determines to suspend or revoke the laboratory's certification, the laboratory's official status will become "Suspended" or "Revoked" until the suspension or revocation is lifted or until any recertification process is complete.

(c) *80 Percent of Participating Laboratories Must Detect Drug.* A laboratory's performance shall be evaluated for all samples for which drugs were spiked at concentrations above the specified performance test level unless the overall response from participating laboratories indicates that less than 80 percent of them were able to detect a drug.

(d) *Participation Required.* Failure to participate in a performance test or to participate satisfactorily may result in suspension or revocation of certification.

3.20 *Inspections.*

Prior to laboratory certification under these Guidelines and at least twice a year after certification, a team of three qualified inspectors, at least two of whom have been trained as laboratory inspectors, shall conduct an on-site inspection of laboratory premises. Inspections shall document the overall quality of the laboratory setting for the purposes of certification to conduct urine drug testing. Inspection reports may also contain recommendations to the laboratory to correct deficiencies noted during the inspection.

3.21 *Results of Inadequate Performance.*

Failure of a laboratory to comply with any aspect of these Guidelines may lead to revocation or suspension of certification as provided in 3.13 and 3.14 of these Guidelines.

[FR Doc. 88-7864 Filed 4-8-88; 8:45 am]
BILLING CODE 4160-20-M

Appendix C

DOT Procedures for Transportation Workplace Drug Testing Programs

PART 40—PROCEDURES FOR TRANSPORTATION WORKPLACE DRUG TESTING PROGRAMS

Sec.
40.1 Applicability.
40.3 Definitions.
40.5–40.19 [Reserved]
40.21 The drugs.
40.23 Preparation for testing.
40.25 Specimen collection procedures.
40.27 Laboratory personnel.
40.29 Laboratory analysis procedures.
40.31 Quality assurance and quality control.
40.33 Reporting and review of results.
40.35 Protection of employee records.
40.37 Individual access to test and laboratory certification results.
40.39 Use of DHHS-certified laboratories.

Appendix A to Part 40—Drug Testing Custody and Control Form

Authority: 49 U.S.C. 102, 301, 322.

§ 40.1 Applicability.

This part applies to transportation employers (including self-employed individuals) conducting drug urine testing programs pursuant to regulations issued by agencies of the Department of Transportation and to such transportation employers' officers, employees, agents and contractors, to the extent and in the manner provided in DOT agency regulations.

§ 40.3 Definitions.

For purposes of this part the following definitions apply:

Aliquot. A portion of a specimen used for testing.

Blind sample or blind performance test specimen. A urine specimen submitted to a laboratory for quality control testing purposes, with a fictitious identifier, so that the laboratory cannot distinguish it from employee specimens, and which is spiked with known quantities of specific drugs or which is blank, containing no drugs.

Chain of custody. Procedures to account for the integrity of each urine specimen by tracking its handling and storage from point of specimen collection to final disposition of the specimen. These procedures shall require that an appropriate drug testing custody form (see § 40.23(a)) be used from time of collection to receipt by the laboratory and that upon receipt by the laboratory an appropriate laboratory chain of custody form(s) account(s) for the sample or sample aliquots within the laboratory.

Collection container. A container into which the employee urinates to provide the urine sample used for a drug test.

Collection site. A place designated by the employer where individuals present themselves for the purpose of providing a specimen of their urine to be analyzed for the presence of drugs.

Collection site person. A person who instructs and assists individuals at a collection site and who receives and makes an initial examination of the urine specimen provided by those individuals.

Confirmatory test. A second analytical procedure to identify the presence of a specific drug or metabolite which is independent of the initial test and which uses a different technique and chemical principle from that of the initial test in order to ensure reliability and accuracy. (Gas chromatography/mass spectrometry (GC/MS) is the only authorized confirmation method for cocaine, marijuana, opiates, amphetamines, and phencyclidine.)

DHHS. The Department of Health and Human Services or any designee of the Secretary, Department of Health and Human Services.

DOT agency. An agency (or "operating administration") of the United States Department of Transportation administering regulations requiring compliance with this part, including the United States Coast Guard, the Federal Aviation Administration, the Federal Railroad Administration, the Federal Highway Administration, the Urban Mass Transportation Administration and the Research and Special Programs Administration.

Employee. An individual designated in a DOT agency regulation as subject to drug urine testing and the donor of a specimen under this part. As used in this part "employee" includes an applicant for employment. "Employee" and "individual" or "individual to be tested" have the same meaning for purposes of this part.

Employer. An entity employing one or more employees that is subject to DOT agency regulations requiring compliance with this part. As used in this part, "employer" includes an industry consortium or joint enterprise comprised of two or more employing entities, but no single employing entity is relieved of its responsibility for compliance with this part by virtue of participation in such a consortium or joint enterprise.

Initial test (also known as screening test). An immunoassay screen to eliminate "negative" urine specimens from further consideration.

Medical Review Officer (MRO). A licensed physician responsible for receiving laboratory results generated by an employer's drug testing program who has knowledge of substance abuse disorders and has appropriate medical training to interpret and evaluate an individual's confirmed positive test result together with his or her medical history and any other relevant biomedical information.

Secretary. The Secretary of Transportation or the Secretary's designee.

Shipping container. A container capable of being secured with a tamper proof seal that is used for transfer of one or more specimen bottle(s) and associated documentation from the collection site to the laboratory.

Specimen bottle. The bottle which, after being labeled and sealed according to the procedures in this part, is used to transmit a urine sample to the laboratory.

§§ 40.5–40.19 [Reserved]

§ 40.21 The drugs.

(a) DOT agency drug testing programs require that employers test for marijuana, cocaine, opiates, amphetamines and phencyclidine.

(b) An employer may include in its testing protocols other controlled substances or alcohol only pursuant to a DOT agency approval, if testing for those substances is authorized under agency regulations and if the DHHS has established an approved testing protocol and positive threshold for each such substance.

(c) Urine specimens collected under DOT agency regulations requiring compliance with this part may only be used to test for controlled substances designated or approved for testing as described in this section and shall not be used to conduct any other analysis or test unless otherwise specifically authorized by DOT agency regulations.

(d) This section does not prohibit procedures reasonably incident to analysis of the specimen for controlled substances (e.g., determination of pH or tests for specific gravity, creatinine concentration or presence of adulterants).

§ 40.23 Preparation for testing.

The employer and certified laboratory shall develop and maintain a clear and well-documented procedure for collection, shipment, and accessioning of urine specimens under this part. Such a procedure shall include, at a minimum, the following:

(a) Utilization of a standard drug testing custody and control form (carbonless manifold). The form shall be a multiple-part, carbonless record form with an original (copy 1), and a "second original" (copy 2), both of which shall accompany the specimen to the laboratory. Copies shall be provided for the Medical Review Officer (copy 3, to go directly to the MRO), the donor (copy

4), the collector (copy 5), and the employer representative (copy 6). If the employer desires to exercise the split sample option, then an additional copy of the urine custody and control form is required. This copy (copy 7) shall be the "split specimen original," and is to accompany the split specimen to the same lab, a second lab, or an employer storage site. There must be a positive link established between the first specimen and the split specimen through the specimen identification number; the split specimen identification number shall be an obvious derivative of the first specimen identification number. The form should be a permanent record on which identifying data on the donor, and on the specimen collection and transfer process, is retained. The form shall be constructed to display, at a minimum, the following elements, which shall appear on its respective parts as indicated:

(1) The following information shall appear on all parts of the form:
(i) A preprinted specimen identification number, which shall be unique to the particular collection. If the split sample option is exercised, the preprinted specimen identification number for split specimen shall be an obvious derivative of the first specimen; e.g., first specimen identification number suffixed "A," split specimen suffixed "B."
(ii) A block specifying the donor's employee identification number or Social Security number, which shall be entered by the collector.
(iii) A block specifying the employer's name, address, and identification number.
(iv) A block specifying the Medical Review Officer's name and address.
(v) Specification for which drugs the specimen identified in this form will be tested.
(vi) Specification for the reason for which this test conducted (preemployment, random, etc.), which shall be entered by the collector.
(vii) A block specifying whether or not the collector read the temperature within 4 minutes, and then notation, by the collector, that the temperature of specimen just read is within the range of 32.5–37.7C/90.5–99.8F; if not within the acceptable range, an area is provided to record the actual temperature.
(viii) A chain-of-custody block providing areas to enter the following information for each transfer of possession: Purpose of change; released by (signature/print name); received by (signature/print name); date. The words "Provide specimen for testing" and "DONOR" shall be preprinted in the initial spaces.

(ix) Information to be completed by the collector: Collector's name; date of collection; location of the collection site: a space for remarks at which unusual circumstances may be described; notation as to whether or not the split specimen was taken in accordance with Federal requirements *if* the option to offer the split specimen was exercised by the employer; and a certification statement as set forth below and a signature block with date which shall be completed by the collector:

I certify that the specimen identified on this form is the specimen presented to me by the donor providing the certification on Copy 3 of this form, that it bears the same identification number as that set forth above, and that it has been collected, labelled and sealed as in accordance with applicable Federal requirements.

(2) Information to be provided by the laboratory after analysis, which shall appear on parts 1, 2 and 7 (if applicable) of the form only: Accession number; laboratory name; address; a space for remarks; specimen results; and certification statement as set forth below, together with spaces to enter the printed name and signature of the certifying laboratory official and date:

I certify that the specimen identified by this accession number is the same specimen that bears the identification number set forth above, that the specimen has been examined upon receipt, handled and analyzed in accordance with applicable Federal requirements, and that the results set forth below are for that specimen.

(3) A block to be completed by the Medical Review Officer (MRO), after the review of the specimen, which shall appear on parts 1, 2 and 7 (if applicable) of the form only, provides for the MRO's name, address, and certification, to read as follows, together with spaces for signature and date:

I have reviewed the laboratory results for the specimen identified by this form in accordance with applicable Federal requirements. My final determination/verification is:

(4) Information to be provided by the donor, which shall appear on parts 3 through 6 of the form only: Donor name (printed); daytime phone number; date of birth; and certification statement as set forth below, together with a signature block with date which shall be completed by the donor.

I certify that I provided my urine specimen to the collector; that the specimen bottle was sealed with a tamper-proof seal in my presence; and that the information provided on this form and on the label affixed to the specimen bottle is correct.

(5) A statement to the donor which shall appear only on parts 3 and 4 of the form, as follows:

Should the results of the laboratory tests for the specimen identified by this form be confirmed positive, the Medical Review Officer will contact you to ask about prescriptions and over-the-counter medications you may have taken. Therefore, you may want to make a list of those medications as a "memory jogger." THIS LIST IS NOT NECESSARY. If you choose to make a list, do so either on a separate piece of paper or on the back of your copy (Copy 4—Donor) of this form—DO NOT LIST ON THE BACK OF ANY OTHER COPY OF THE FORM. TAKE YOUR COPY WITH YOU.

A form meeting the requirements of this paragraph is displayed at appendix A to this part.

(6) The drug testing custody and control form may include such additional information as may be required for billing or other legitimate purposes necessary to the collection, provided that personal identifying information on the donor (other than the social security number) may not be provided to the laboratory. Donor medical information may appear only on the copy provided to the donor.

(b)(1) Use of a clean, single-use specimen bottle that is securely wrapped until filled with the specimen. A clean, single-use collection container (e.g., disposable cup or sterile urinal) that is securely wrapped until used may also be employed. *If urination is directly into the specimen bottle,* the specimen bottle shall be provided to the employee still sealed in its wrapper or shall be unwrapped in the employee's presence immediately prior to its being provided. *If a separate collection container is used for urination,* the collection container shall be provided to the employee still sealed in its wrapper or shall be unwrapped in the employee's presence immediately prior to its being provided; and the collection site person shall unwrap the specimen bottle in the presence of the employee at the time the urine specimen is presented.

(2) Use of a tamperproof sealing system, designed in a manner such to ensure against undetected opening. The specimen bottle shall be identified with a unique identifying number identical to that appearing on the urine custody and control form, and space shall be provided to initial the bottle affirming its identity. For purposes of clarity, this part assumes use of a system made up of one or more preprinted labels and seals (or a unitary label/seal), but use of other, equally effective technologies is authorized.

DOT Procedures 221

(c) Use of a shipping container in which the specimen and associated paperwork may be transferred and which can be sealed and initialled to prevent undetected tampering. In the split specimen option is exercised, the split specimen and associated paperwork shall be sealed in a shipping (or storage) container and initialled to prevent undetected tampering.

(d) Written procedures, instructions and training shall be provided as follows:

(1) Employer collection procedures and training shall clearly emphasize that the collection site person is responsible for maintaining the integrity of the specimen collection and transfer process, carefully ensuring the modesty and privacy of the donor, and is to avoid any conduct or remarks that might be construed as accusatorial or otherwise offensive or inappropriate.

(2) A collection site person shall have successfully completed training to carry out this function or shall be a licensed medical professional or technician who is provided instructions for collection under this part and certifies completion as required in this part

(i) A non-medical collection site person shall receive training in compliance with this part and shall demonstrate proficiency in the application of this part prior to serving as a collection site person. A medical professional, technologist or technician licensed or otherwise approved to practice in the jurisdiction in which the collection takes place is not required to receive such training if that person is provided instructions described in this part and performs collections in accordance with those instructions.

(ii) Collection site persons shall be provided with detailed, clear instructions on the collection of specimens in compliance with this part. Employer representatives and donors subject to testing shall also be provided standard written instructions setting forth their responsibilities.

(3) Unless it is impracticable for any other individual to perform this function, a direct supervisor of an employee shall not serve as the collection site person for a test of the employee. If the rules of a DOT agency are more stringent than this provision regarding the use of supervisors as collection site personnel, the DOT agency rules shall prevail with respect to testing to which they apply.

(4) In any case where a collection is monitored by non-medical personnel or is directly observed, the collection site person shall be of the same gender as the donor. A collection is monitored for this purpose if the enclosure provides less than complete privacy for the donor (e.g., if a restroom stall is used and the collection site person remains in the restroom, or if the collection site person is expected to listen for use of unsecured sources of water.)

§ 40.25 Specimen collection procedures.

(a) *Designation of collection site.* (1) Each employer drug testing program shall have one or more designated collection sites which have all necessary personnel, materials, equipment, facilities and supervision to provide for the collection, security, temporary storage, and shipping or transportation of uring specimens to a certified drug testing laboratory. An independent medical facility may also be utilized as a collection site provided the other applicable requirements of this part are met.

(2) A designated collection site may be any suitable location where a specimen can be collected under conditions set forth in this part, including a properly equipped mobile facility. A designated collection site shall be a location having an enclosure within which private urination can occur, a toilet for completion of urination (unless a single-use collector is used with sufficient capacity to contain the void), and a suitable clean surface for writing. The site must also have a source of water for washing hands, which, if practicable, should be external to the enclosure where urination occurs.

(b) *Security.* The purpose of this paragraph is to prevent unauthorized access which could compromise the integrity of the collection process or the specimen.

(1) Procedures shall provide for the designated collection site to be secure. If a collection site facility is dedicated solely to urine collection, it shall be secure at all times. If a facility cannot be dedicated solely to drug testing, the portion of the facility used for testing shall be secured during drug testing.

(2) A facility normally used for other purposes, such as a public rest room or hospital examining room, may be secured by visual inspection to ensure other persons are not present and undetected access (e.g., through a rear door not in the view of the collection site person) is not possible. Security during collection may be maintained by effective restriction of access to collection materials and specimens. In the case of a public rest room, the facility must be posted against access during the entire collection procedure to avoid embarrassment to the employee or distraction of the collection site person.

(3) If it is impractical to maintain continuous physical security of a collection site from the time the specimen is presented until the sealed mailer is transferred for shipment, the following minimum procedures shall apply. The specimen shall remain under the direct control of the collection site person from delivery to its being sealed in the mailer. The mailer shall be immediately mailed, maintained in secure storage, or remain until mailed under the personal control of the collection site person.

(c) *Chain of custody.* The chain of custody block of the drug testing custody and control form shall be properly executed by authorized collection site personnel upon receipt of specimens. Handling and transportation of urine specimens from one authorized individual or place to another shall always be accomplished through chain of custody procedures. Every effort shall be made to minimize the number of persons handling specimens.

(d) *Access to authorized personnel only.* No unauthorized personnel shall be permitted in any part of the designated collection site where urine specimens are collected or stored. Only the collection site person may handle specimens prior to their securement in the mailing container or monitor or observe specimen collection (under the conditions specified in this part). In order to promote security of specimens, avoid distraction of the collection site person and ensure against any confusion in the identification of specimens, the collection site person shall have only one donor under his or her supervision at any time. For this purpose, a collection procedure is complete when the urine bottle has been sealed and initialled, the drug testing custody and control form has been executed, and the employee has departed the site (or, in the case of an employee who was unable to provide a complete specimen, has entered a waiting area).

(e) *Privacy.* (1) Procedures for collecting urine specimens shall allow individual privacy unless there is a reason to believe that a particular individual may alter or substitute the specimen to be provided, as further described in this paragraph.

(2) For purposes of this part, the following circumstances are the exclusive grounds constituting a reason to believe that the individual may alter or substitute the specimen:

(i) The employee has presented a urine specimen that falls outside the normal temperature range (32.5°–37.7 °C/90.5°–99.8 °F), and

(A) The employee declines to provide a measurement of oral body temperature, as provided in paragraph (f)(14) of the part; or

(B) Oral body temperature varies by more than 1°C/1.8°F from the temperature of the specimen;

(ii) The last urine specimen provided by the employee (i.e., on a previous occasion) was determined by the laboratory to have a specific gravity of less than 1.003 and a creatinine concentration below .2g/L;

(iii) The collection site person observes conduct clearly and unequivocally indicating an attempt to substitute or adulterate the sample (e.g., substitute urine in plain view, blue dye in specimen presented, etc.); or

(iv) The employee has previously been determined to have used a controlled substance without medical authorization and the particular test was being conducted under a DOT agency regulation providing for follow-up testing upon or after return to service.

(3) A higher-level supervisor of the collection site person, or a designated employer representative, shall review and concur in advance with any decision by a collection site person to obtain a specimen under the direct observation of a same gender collection site person based upon the circumstances described in subparagraph (2) of this paragraph.

(f) *Integrity and identity of specimen.* Employers shall take precautions to ensure that a urine specimen is not adulterated or diluted during the collection procedure and that information on the urine bottle and on the urine custody and control form can identify the individual from whom the specimen was collected. The following minimum precautions shall be taken to ensure that unadulterated specimens are obtained and correctly identified:

(1) To deter the dilution of specimens at the collection site, toilet bluing agents shall be placed in toilet tanks wherever possible, so the reservoir of water in the toilet bowl always remains blue. Where practicable, there shall be no other source of water (e.g., shower or sink) in the enclosure where urination occurs. If there is another source of water in the enclosure it shall be effectively secured or monitored to ensure it is not used as a source for diluting the specimen.

(2) When an individual arrives at the collection site, the collection site person shall ensure that the individual is positively identified as the employee selected for testing (e.g., through presentation of photo identification or identification by the employer's representative). If the individual's identity cannot be established, the collection site person shall not proceed with the collection. If the employee requests, the collection site person shall show his/her identification to the employee.

(3) If the individual fails to arrive at the assigned time, the collection site person shall contact the appropriate authority to obtain guidance on the action to be taken.

(4) The collection site person shall ask the individual to remove any unnecessary outer garments such as a coat or jacket that might conceal items or substances that could be used to tamper with or adulterate the individual's urine specimen. The collection site person shall ensure that all personal belongings such as a purse or briefcase remain with the outer garments. The individual may retain his or her wallet. If the employee requests it, the collection site personnel shall provide the employee a receipt for any personal belongings.

(5) The individual shall be instructed to wash and dry his or her hands prior to urination.

(6) After washing hands, the individual shall remain in the presence of the collection site person and shall not have access to any water fountain, faucet, soap dispenser, cleaning agent or any other materials which could be used to adulterate the specimen.

(7) The individual may provide his/her specimen in the privacy of a stall or otherwise partitioned area that allows for individual privacy. The collection site person shall provide the individual with a specimen bottle or collection container, if applicable, for this purpose.

(8) The collection site person shall note any unusual behavior or appearance on the urine custody and control form.

(9) In the exceptional event that an employer-designated collection site is not accessible and there is an immediate requirement for specimen collection (e.g., circumstances require a post-accident test), a public rest room may be used according to the following procedures: A collection site person of the same gender as the individual shall accompany the individual into the public rest room which shall be made secure during the collection procedure. If possible, a toilet bluing agent shall be placed in the bowl and any accessible toilet tank. The collection site person shall remain in the rest room, but outside the stall, until the specimen is collected. If no bluing agent is available to deter specimen dilution, the collection site person shall instruct the individual not to flush the toilet until the specimen is delivered to the collection site person. After the collection site person has possession of the specimen, the individual will be instructed to flush the toilet and to participate with the collection site person in completing the chain of custody procedures.

(10)(i) Upon receiving the specimen from the individual, the collection site person shall determine if it contains at least 60 milliliters of urine. If the individual is unable to provide a 60 milliliters of urine, the collection site person shall direct the individual to drink fluids and, after a reasonable time, again attempt to provide a complete sample using a fresh specimen bottle (and fresh collection container, if employed). The original specimen shall be discarded. If the employee is still unable to provide a complete specimen, the following rules apply:

(A) In the case of a post-accident test or test for reasonable cause (as defined by the DOT agency), the employee shall remain at the collection site and continue to consume reasonable quantities of fluids until the specimen has been provided or until the expiration of a period up to 8 hours from the beginning of the collection procedure.

(B) In the case of a preemployment test, random test, periodic test or other test not for cause (as defined by the DOT agency), the employer may elect to proceed as specified in paragraph (f)(10)(i)(A) of this section (consistent with any applicable restrictions on hours of service) or may elect to discontinue the collection and conduct a subsequent collection at a later time.

(C) If the employee cannot provide a complete sample within the up to 8-hour period or at the subsequent collection, as applicable, then the employer's MRO shall refer the individual for a medical evaluation to develop pertinent information concerning whether the individual's inability to provide a specimen is genuine or constitutes a refusal to provide a specimen. (In preemployment testing, if the employer does not wish to hire the individual, the MRO is not required to make such a referral.) Upon completion of the examination, the MRO shall report his or her conclusions to the employer in writing.

(ii) The employer may, but is not required to, use a "split sample" method of collection.

(A) The donor shall urinate into a collection container, which the collection site person, in the presence of the donor, after determining specimen temperature, pours into two specimen bottles.

(B) The first bottle is to be used for the DOT-mandated test, and 60 ml of urine shall be poured into it. If there is no additional urine available for the second specimen bottle, the first specimen bottle shall nevertheless be processed for testing.

(C) Up to 60 ml of the remainder of the urine shall be poured into the second specimen bottle.

(D) All requirements of this part shall be followed with respect to both samples, including the requirement that a copy of the chain of custody form accompany each bottle processed under "split sample" procedures.

(E) Any specimen collected under "split sample" procedures must be stored in a secured, refrigerated environment and an appropriate entry made in the chain of custody form.

(F) If the test of the first bottle is positive, the employee may request that the MRO direct that the second bottle be tested in a DHHS-certified laboratory for presence of the drug(s) for which a positive result was obtained in the test of the first bottle. The result of this test is transmitted to the MRO without regard to the cutoff values of § 40.29. The MRO shall honor such a request if it is made within 72 hours of the employee's having actual notice that he or she tested positive.

(G) Action required by DOT regulations as the result of a positive drug test (e.g., removal from performing a safety-sensitive function) is not stayed pending the result of the second test.

(H) If the result of the second test is negative, the MRO shall cancel the test.

(11) After the specimen has been provided and submitted to the collection site person, the individual shall be allowed to wash his or her hands.

(12) Immediately after the specimen is collected, the collection site person shall measure the temperature of the specimen. The temperature measuring device used must accurately reflect the temperature of the specimen and not contaminate the specimen. The time from urination to temperature measure is critical and in no case shall exceed 4 minutes.

(13) A specimen temperature outside the range of 32.5°–37.7 °C/90.5°–99.8 °F constitutes a reason to believe that the individual has altered or substituted the specimen (see paragraph (e)(2)(i) of this section). In such cases, the individual supplying the specimen may volunteer to have his or her oral temperature taken to provide evidence to counter the reason to believe the individual may have altered or substituted the specimen.

(14) Immediately after the specimen is collected, the collection site person shall also inspect the specimen to determine its color and look for any signs of contaminants. Any unusual findings shall be noted on the urine custody and control form.

(15) All specimens suspected of being adulterated shall be forwarded to the laboratory for testing.

(16) Whenever there is reason to believe that a particular individual has altered or substituted the specimen as described in paragraph (e)(2) (i) or (iii) of this section, a second specimen shall be obtained as soon as possible under the direct observation of a same gender collection site person.

(17) Both the individual being tested and the collection site person shall keep the specimen in view at all times prior to its being sealed and labeled. As provided below, the specimen shall be sealed (by placement of a tamperproof seal over the bottle cap and down the sides of the bottle) and labeled in the presence of the employee. If the specimen is transferred to a second bottle, the collection site person shall request the individual to observe the transfer of the specimen and the placement of the tamperproof seal over the bottle cap and down the sides of the bottle.

(18) The collection site person and the individual being tested shall be present at the same time during procedures outlined in paragraphs (f)(19)–(f)(22) of this section.

(19) The collection site person shall place securely on the bottle an identification label which contains the date, the individual's specimen number, and any other identifying information provided or required by the employer. If separate from the label, the tamperproof seal shall also be applied.

(20) The individual shall initial the identification label on the specimen bottle for the purpose of certifying that it is the specimen collected from him or her.

(21) The collection site person shall enter on the drug testing custody and control form all information identifying the specimen. The collection site person shall sign the drug testing custody and control form certifying that the collection was accomplished according to the applicable Federal requirements.

(22)(i) The individual shall be asked to read and sign a statement on the drug testing custody and control form certifying that the specimen identified as having been collected from him or her is in fact the specimen he or she provided.

(ii) When specified by DOT agency regulation or required by the collection site (other than an employer site) or by the laboratory, the employee may be required to sign a consent or release form authorizing the collection of the specimen, analysis of the specimen for designated controlled substances, and release of the results to the employer. The employee may not be required to waive liability with respect to negligence on the part of any person participating in the collection, handling or analysis of the specimen or to indemnify any person for the negligence of others.

(23) The collection site person shall complete the chain of custody portion of the drug testing custody and control form to indicate receipt of the specimen from the employee and shall certify proper completion of the collection.

(24) The urine specimen and chain of custody form are now ready for shipment. If the specimen is not immediately prepared for shipment, the collection site person shall ensure that it is appropriately safeguarded during temporary storage.

(25)(i) While any part of the above chain of custody procedures is being performed, it is essential that the urine specimen and custody documents be under the control of the involved collection site person. If the involved collection site person leaves his or her work station momentarily, the collection site person shall take the specimen and drug testing custody and control form with him or her or shall secure them. After the collection site person returns to the work station, the custody process will continue. If the collection site person is leaving for an extended period of time, he or she shall package the specimen for mailing before leaving the site.

(ii) The collection site person shall not leave the collection site in the interval between presentation of the specimen by the employee and securement of the sample with an identifying label bearing the employee's specimen identification number (shown on the urine custody and control form) and seal initialed by the employee. If it becomes necessary for the collection site person to leave the site during this interval, the collection shall be nullified and (at the election of the employer) a new collection begun.

(g) *Collection control.* To the maximum extent possible, collection site personnel shall keep the individual's specimen bottle within sight both before and after the individual has urinated. After the specimen is collected, it shall be properly sealed and labeled.

(h) *Transportation to laboratory.* Collection site personnel shall arrange to ship the collected specimen to the drug testing laboratory. The specimens shall be placed in shipping containers designed to minimize the possibility of damage during shipment (e.g., specimen boxes and/or padded mailers); and those containers shall be securely sealed to eliminate the possibility of undetected tampering. On the tape sealing the container, the collection site person shall sign and enter the date specimens were sealed in the shipping containers for shipment. The collection site person shall ensure that the chain of custody documentation is attached or enclosed in each container sealed for shipment to the drug testing laboratory.

(i) *Failure to cooperate.* If the employee refuses to cooperate with the collection process, the collection site person shall inform the employer representative and shall document the non-cooperation on the drug testing custody and control form.

(j) *Employee requiring medical attention.* If the sample is being collected from an employee in need of medical attention (e.g., as part of a post-accident test given in an emergency medical facility), necessary medical attention shall not be delayed in order to collect the specimen.

(k) *Use of chain of custody forms.* A chain of custody form (and a laboratory internal chain of custody document, where applicable) shall be used for maintaining control and accountability of each specimen from the point of collection to final disposition of the specimen. The date and purpose shall be documented on the form each time a specimen is handled or transferred and every individual in the chain shall be identified. Every effort shall be made to minimize the number of persons handling specimens.

§ 40.27 Laboratory personnel.

(a) *Day-to-day management.* (1) The laboratory shall have a qualified individual to assume professional, organizational, educational, and administrative responsibility for the laboratory's urine drug testing facility.

(2) This individual shall have documented scientific qualifications in analytical forensic toxicology. Minimum qualifications are:

(i) Certification as a laboratory director by a State in forensic or clinical laboratory toxicology; or

(ii) A Ph.D. in one of the natural sciences with an adequate undergraduate and graduate education in biology, chemistry, and pharmacology or toxicology; or

(iii) Training and experience comparable to a Ph.D. in one of the natural sciences, such as a medical or scientific degree with additional training and laboratory/research experience in biology, chemistry, and pharmacology or toxicology; and

(iv) In addition to the requirements in paragraph (a)(2) (i), (ii), or (iii) of this section, minimum qualifications also require:

(A) Appropriate experience in analytical forensic toxicology including experience with the analysis of biological material for drugs of abuse, and

(B) Appropriate training and/or experience in forensic applications of analytical toxicology, e.g., publications, court testimony, research concerning analytical toxicology of drugs of abuse, or other factors which qualify the individual as an expert witness in forensic toxicology.

(3) This individual shall be engaged in and responsible for the day-to-day management of the drug testing laboratory even where another individual has overall responsibility for an entire multi-specialty laboratory.

(4) This individual shall be responsible for ensuring that there are enough personnel with adequate training and experience to supervise and conduct the work of the drug testing laboratory. He or she shall assure the continued competency of laboratory personnel by documenting their in-service training, reviewing their work performance, and verifying their skills.

(5) This individual shall be responsible for the laboratory's having a procedure manual which is complete, up-to-date, available for personnel performing tests, and followed by those personnel. The procedure manual shall be reviewed, signed, and dated by this responsible individual whenever procedures are first placed into use or changed or when a new individual assumes responsibility for management of the drug testing laboratory. Copies of all procedures and dates on which they are in effect shall be maintained. (Specific contents of the procedure manual are described in § 40.29(n)(1).)

(6) This individual shall be responsible for maintaining a quality assurance program to assure the proper performance and reporting of all test results; for maintaining acceptable analytical performance for all controls and standards; for maintaining quality control testing; and for assuring and documenting the validity, reliability, accuracy, precision, and performance characteristics of each test and test system.

(7) This individual shall be responsible for taking all remedial actions necessary to maintain satisfactory operation and performance of the laboratory in response to quality control systems not being within performance specifications, errors in result reporting or in analysis of performance testing results. This individual shall ensure that sample results are not reported until all corrective actions have been taken and he or she can assure that the tests results provided are accurate and reliable.

(b) *Test validation.* The laboratory's urine drug testing facility shall have a qualified individual(s) who reviews all pertinent data and quality control results in order to attest to the validity of the laboratory's test reports. A laboratory may designate more than one person to perform this function. This individual(s) may be any employee who is qualified to be responsible for day-to-day management or operation of the drug testing laboratory.

(c) *Day-to-day operations and supervision of analysts.* The laboratory's urine drug testing facility shall have an individual to be responsible for day-to-day operations and to supervise the technical analysts. This individual(s) shall have at least a bachelor's degree in the chemical or biological sciences or medical technology or equivalent. He or she shall have training and experience in the theory and practice of the procedures used in the laboratory, resulting in his or her thorough understanding of quality control practices and procedures; the review, interpretation, and reporting of test results; maintenance of chain of custody; and proper remedial actions to be taken in response to test systems being out of control limits or detecting aberrant test or quality control results.

(d) *Other personnel.* Other technicians or nontechnical staff shall have the necessary training and skills for the tasks assigned.

(e) *Training.* The laboratory's urine drug testing program shall make available continuing education programs to meet the needs of laboratory personnel.

(f) *Files.* Laboratory personnel files shall include: resume of training and experience, certification or license if any; references; job descriptions; records of performance evaluation and advancement; incident reports; and results of tests which establish employee competency for the position he or she holds, such as a test for color blindness, if appropriate.

§ 40.29 Laboratory analysis procedures.

(a) *Security and chain of custody.* (1) Drug testing laboratories shall be secure at all times. They shall have in place sufficient security measures to control access to the premises and to ensure that no unauthorized personnel handle specimens or gain access to the laboratory process or to areas where records are stored. Access to these secured areas shall be limited to specifically authorized individuals whose authorization is documented. With the exception of personnel authorized to conduct inspections on behalf of Federal agencies for which the laboratory is engaged in urine testing or on behalf of DHHS, all authorized visitors and maintenance and service personnel shall be escorted at all times. Documentation of individuals accessing these areas, dates, and time of entry and purpose of entry must be maintained.

(2) Laboratories shall use chain of custody procedures to maintain control and accountability of specimens from receipt through completion of testing, reporting of results during storage, and continuing until final disposition of specimens. The date and purpose shall be documented on an appropriate chain of custody form each time a specimen is handled or transferred and every individual in the chain shall be identified. Accordingly, authorized technicians shall be responsible for each urine specimen or aliquot in their possession and shall sign and complete chain of custody forms for those specimens or aliquots as they are received.

(b) *Receiving.* (1) When a shipment of specimens is received, laboratory personnel shall inspect each package for evidence of possible tampering and compare information on specimen bottles within each package to the information on the accompanying chain of custody forms. Any direct evidence of tampering or discrepancies in the information on specimen bottles and the employer's chain of custody forms attached to the shipment shall be immediately reported to the employer and shall be noted on the laboratory's chain of custody form which shall accompany the specimens while they are in the laboratory's possession.

(2) Specimen bottles generally shall be retained within the laboratory's accession area until all analyses have been completed. Aliquots and the laboratory's chain of custody forms shall be used by laboratory personnel for conducting initial and confirmatory tests.

(c) *Short-term refrigerated storage.* Specimens that do not receive an initial test within 7 days of arrival at the laboratory shall be placed in secure refrigeration units. Temperatures shall not exceed 6°C. Emergency power equipment shall be available in case of prolonged power failure.

(d) *Specimen processing.* Laboratory facilities for urine drug testing will normally process specimens by grouping them into batches. The number of specimens in each batch may vary significantly depending on the size of the laboratory and its workload. When conducting either initial or confirmatory tests, every batch shall contain an appropriate number of standards for calibrating the instrumentation and a minimum of 10 percent controls. Both quality control and blind performance test samples shall appear as ordinary samples to laboratory analysts.

(e) *Initial test.* (1) The initial test shall use an immunoassay which meets the requirements of the Food and Drug Administration for commercial distribution. The following initial cutoff levels shall be used when screening specimens to determine whether they are negative for these five drugs or classes of drugs:

	Initial test cutoff levels (ng/ml)
Marijuana metabolites	100
Cocaine metabolites	300
Opiate metabolites	*300
Phencyclidine	25
Amphetamines	1,000

*25 ng/ml if immunoassay specific for free morphine.

(2) These cutoff levels are subject to change by the Department of Health and Human Services as advances in technology or other considerations warrant identification of these substances at other concentrations.

(f) *Confirmatory test.* (1) All specimens identified as positive on the initial test shall be confirmed using gas chromatography/mass spectrometry (GC/MS) techniques at the cutoff levels listed in this paragraph for each drug. All confirmations shall be by quantitative analysis. Concentrations that exceed the linear region of the standard curve shall be documented in the laboratory record as "greater than highest standard curve value."

	Confirmatory test cutoff levels (ng/ml)
Marijuana metabolite[a]	15
Cocaine metabolite[b]	150
Opiates:	
Morphine	300
Codeine	300
Phencyclidine	25
Amphetamines:	
Amphetamine	500
Methamphetamine	500

[a] Delta-9-tetrahydrocannabinol-9-carboxylic acid.
[b] Benzoylecgonine.

(2) These cutoff levels are subject to change by the Department of Health and Human Services as advances in technology or other considerations warrant identification of these substances at other concentrations.

(g) *Reporting results.* (1) The laboratory shall report test results to the employer's Medical Review Officer within an average of 5 working days after receipt of the specimen by the laboratory. Before any test result is reported (the results of initial tests, confirmatory tests, or quality control data), it shall be reviewed and the test certified as an accurate report by the responsible individual. The report shall identify the drugs/metabolites tested for, whether positive or negative, the specimen number assigned by the employer, and the drug testing laboratory specimen identification number (accession number).

(2) The laboratory shall report as negative all specimens that are negative on the initial test or negative on the confirmatory test. Only specimens confirmed positive shall be reported positive for a specific drug.

(3) The Medical Review Officer may request from the laboratory and the laboratory shall provide quantitation of test results. The MRO shall report whether the test is positive or negative, and may report the drug(s) for which there was a positive test, but shall not disclose the quantitation of test results to the employer. *Provided,* that the MRO may reveal the quantitation of a positive test result to the employer, the employee, or the decisionmaker in a lawsuit, grievance, or other proceeding initiated by or on behalf of the employee and arising from a verified positive drug test.

(4) The laboratory may transmit results to the Medical Review Officer by various electronic means (for example, teleprinters, facsimile, or computer) in a manner designed to ensure confidentiality of the information. Results may not be provided verbally by telephone. The laboratory and employer must ensure the security of the data transmission and limit access to any data transmission, storage, and retrieval system.

(5) The laboratory shall send only to the Medical Review Officer the original or a certified true copy of the drug testing custody and control form (part 2), which, in the case of a report positive for drug use, shall be signed (after the required certification block) by the individual responsible for day-to-day management of the drug testing laboratory or the individual responsible for attesting to the validity of the test reports, and attached to which shall be a copy of the test report.

(6) The laboratory shall provide to the employer official responsible for coordination of the drug testing program a monthly statistical summary of

urinalysis testing of the employer's employees and shall not include in the summary any personal identifying information. Initial and confirmation data shall be included from test results reported within that month. Normally this summary shall be forwarded by registered or certified mail not more than 14 calendar days after the end of the month covered by the summary. The summary shall contain the following information:

(i) Initial Testing:
(A) Number of specimens received;
(B) Number of specimens reported out; and
(C) Number of specimens screened positive for:
Marijuana metabolites
Cocaine metabolites
Opiate metabolites
Phencyclidine
Amphetamine
(ii) Confirmatory Testing:
(A) Number of specimens received for confirmation;
(B) Number of specimens confirmed positive for:
Marijuana metabolite
Cocaine metabolite
Morphine, codeine
Phencyclidine
Amphetamine
Methamphetamine

Monthly reports shall not include data from which it is reasonably likely that information about individuals' tests can be readily inferred. If necessary, in order to prevent the disclosure of such data, the laboratory shall not send a report until data are sufficiently aggregated to make such an inference unlikely. In any month in which a report is withheld for this reason, the laboratory will so inform the employer in writing.

(7) The laboratory shall make available copies of all analytical results for employer drug testing programs when requested by DOT or any DOT agency with regulatory authority over the employer.

(8) Unless otherwise instructed by the employer in writing, all records pertaining to a given urine specimen shall be retained by the drug testing laboratory for a minimum of 2 years.

(h) *Long-term storage.* Long-term frozen storage ($-20°C$ or less) ensures that positive urine specimens will be available for any necessary retest during administrative or disciplinary proceedings. Drug testing laboratories shall retain and place in properly secured long-term frozen storage for a minimum of 1 year all specimens confirmed positive, in their original labeled specimen bottles. Within this 1-year period, an employer (or other person designated in a DOT agency regulation) may request the laboratory to retain the specimen for an additional period of time, but if no such request is received the laboratory may discard the specimen after the end of 1 year, except that the laboratory shall be required to maintain any specimens known to be under legal challenge for an indefinite period.

(i) *Retesting specimens.* Because some analytes deteriorate or are lost during freezing and/or storage, quantitation for a retest is not subject to a specific cutoff requirement but must provide data sufficient to confirm the presence of the drug or metabolite.

(j) *Subcontracting.* Drug testing laboratories shall not subcontract and shall perform all work with their own personnel and equipment. The laboratory must be capable of performing testing for the five classes of drugs (marijuana, cocaine, opiates, phencyclidine and amphetamines) using the initial immunoassay and confirmatory GC/MS methods specified in this part. This paragraph does not prohibit subcontracting of laboratory analysis if specimens are sent directly from the collection site to the subcontractor, the subcontractor is a laboratory certified by DHHS as required in this part, the subcontractor performs all analysis and provides storage required under this part, and the subcontractor is responsible to the employer for compliance with this part and applicable DOT agency regulations as if it were the prime contractor.

(k) *Laboratory facilities.* (1) Laboratory facilities shall comply with applicable provisions of any State licensing requirements.

(2) Laboratories certified in accordance with DHHS Guidelines shall have the capability, at the same laboratory premises, of performing initial and confirmatory tests for each drug or metabolite for which service is offered.

(l) *Inspections.* The Secretary, a DOT agency, any employer utilizing the laboratory, DHHS or any organization performing laboratory certification on behalf of DHHS reserves the right to inspect the laboratory at any time. Employer contracts with laboratories for drug testing, as well as contracts for collection site services, shall permit the employer and the DOT agency of jurisdiction (directly or through an agent) to conduct unannounced inspections.

(m) *Documentation.* The drug testing laboratories shall maintain and make available for at least 2 years documentation of all aspects of the testing process. This 2 year period may be extended upon written notification by a DOT agency or by any employer for which laboratory services are being provided. The required documentation shall include personnel files on all individuals authorized to have access to specimens; chain of custody documents; quality assurance/quality control records; procedure manuals; all test data (including calibration curves and any calculations used in determining test results); reports; performance records on performance testing; performance on certification inspections; and hard copies of computer-generated data. The laboratory shall maintain documents for any specimen known to be under legal challenge for an indefinite period.

(n) *Additional requirements for certified laboratories.*—(1) *Procedure manual.* Each laboratory shall have a procedure manual which includes the principles of each test preparation of reagents, standards and controls, calibration procedures, derivation of results, linearity of methods, sensitivity of methods, cutoff values, mechanisms for reporting results, controls criteria for unacceptable specimens and results, remedial actions to be taken when the test systems are outside of acceptable limits, reagents and expiration dates, and references. Copies of all procedures and dates on which they are in effect shall be maintained as part of the manual.

(2) *Standards and controls.* Laboratory standards shall be prepared with pure drug standards which are properly labeled as to content and concentration. The standards shall be labeled with the following dates: when received; when prepared or opened; when placed in service; and expiration date.

(3) *Instruments and equipment.* (i) Volumetric pipettes and measuring devices shall be certified for accuracy or be checked by gravimetric, colorimetric, or other verification procedure. Automatic pipettes and dilutors shall be checked for accuracy and reproducibility before being placed in service and checked periodically thereafter.

(ii) There shall be written procedures for instrument set-up and normal operation, a schedule for checking critical operating characteristics for all instruments, tolerance limits for acceptable function checks and instructions for major trouble shooting and repair. Records shall be available on preventive maintenance.

(4) *Remedial actions.* There shall be written procedures for the actions to be taken when systems are out of acceptable limits or errors are detected. There shall be documentation that these

procedures are followed and that all necessary corrective actions are taken. There shall also be in place systems to verify all stages of testing and reporting and documentation that these procedures are followed.

(5) *Personnel available to testify at proceedings.* A laboratory shall have qualified personnel available to testify in an administrative or disciplinary proceeding against an employee when that proceeding is based on positive urinalysis results reported by the laboratory.

§ 40.31 Quality assurance and quality control.

(a) *General.* Drug testing laboratories shall have a quality assurance program which encompasses all aspects of the testing process including but not limited to specimen acquisition, chain of custody security and reporting of results, initial and confirmatory testing and validation of analytical procedures. Quality assurance procedures shall be designed, implemented and reviewed to monitor the conduct of each step of the process of testing for drugs.

(b) *Laboratory quality control requirements for initial tests.* Each analytical run of specimens to be screened shall include:

(1) Urine specimens certified to contain no drug;

(2) Urine specimens fortified with known standards; and

(3) Positive controls with the drug or metabolite at or near the cutoff level.

In addition, with each batch of samples a sufficient number of standards shall be included to ensure and document the linearity of the assay method over time in the concentration area of the cutoff. After acceptable values are obtained for the known standards, those values will be used to calculate sample data. Implementation of procedures to ensure the carryover does not contaminate the testing of an individual's specimen shall be documented. A minimum of 10 percent of all test samples shall be quality control specimens. Laboratory quality control samples, prepared from spiked urine samples of determined concentration shall be included in the run and should appear as normal samples to laboratory analysts. One percent of each run, with a minimum of at least one sample, shall be the laboratory's own quality control samples.

(c) *Laboratory quality control requirements for confirmation tests.* Each analytical run of specimens to be confirmed shall include:

(1) Urine specimens certified to contain no drug;

(2) Urine specimens fortified with known standards; and

(3) Positive controls with the drug or metabolite at or near the cutoff level. The linearity and precision of the method shall be periodically documented. Implementation of procedures to ensure that carryover does not contaminate the testing of an individual's specimen shall also be documented.

(d) *Employer blind performance test procedures.*

(1) Each employer covered by DOT agency drug testing regulations shall use blind testing quality control procedures as provided in this paragraph.

(2) Each employer shall submit three blind performance test specimens for each 100 employee specimens it submits, up to a maximum of 100 blind performance test specimens submitted per quarter. A DOT agency may increase this per quarter maximum number of samples if doing so is necessary to ensure adequate quality control of employers or consortiums with very large numbers of employees.

(3) For employers with 2000 or more covered employees, approximately 80 percent of the blind performance test samples shall be blank (i.e., containing no drug or otherwise as approved by a DOT agency) and the remaining samples shall be positive for one or more drugs per sample in a distribution such that all the drugs to be tested are included in approximately equal frequencies of challenge. The positive samples shall be spiked only with those drugs for which the employer is testing. This paragraph shall not be construed to prohibit spiking of other (potentially interfering) compounds, as technically appropriate, in order to verify the specificity of a particular assay.

(4) Employers with fewer than 2000 covered employees may submit blind performance test specimens as provided in paragraph (d)(3) of this section. Such employers may also submit only blank samples or may submit two separately labeled portions of a specimen from the same non-covered employee.

(5) Consortiums shall be responsible for the submission of blind samples on behalf of their members. The blind sampling rate shall apply to the total number of samples submitted by the consortium.

(6) The DOT agency concerned shall investigate, or shall refer to DHHS for investigation, any unsatisfactory performance testing result and, based on this investigation, the laboratory shall take action to correct the cause of the unsatisfactory performance test result. A record shall be made of the investigative findings and the corrective action taken by the laboratory, and that record shall be dated and signed by the individual responsible for the day-to-day management and operation of the drug testing laboratory. Then the DOT agency shall send the document to the employer as a report of the unsatisfactory performance testing incident. The DOT agency shall ensure notification of the finding to DHHS.

(7) Should a false positive error occur on a blind performance test specimen and the error is determined to be an administrative error (clerical, sample mixup, etc.), the employer shall promptly notify the DOT agency concerned. The DOT agency and the employer shall require the laboratory to take corrective action to minimize the occurrence of the particular error in the future, and, if there is reason to believe the error could have been systemic, the DOT agency may also require review and reanalysis of previously run specimens.

(8) Should a false positive error occur on a blind performance test specimen and the error is determined to be a technical or methodological error, the employer shall instruct the laboratory to submit all quality control data from the batch of specimens which included the false positive specimen to the DOT agency concerned. In addition, the laboratory shall retest all specimens analyzed positive for that drug or metabolite from the time of final resolution of the error back to the time of the last satisfactory performance test cycle. This retesting shall be documented by a statement signed by the individual responsible for day-to-day management of the laboratory's urine drug testing. The DOT agency concerned may require an on-site review of the laboratory which may be conducted unannounced during any hours of operation of the laboratory. Based on information provided by the DOT agency, DHHS has the option of revoking or suspending the laboratory's certification or recommending that no further action be taken if the case is one of less serious error in which corrective action has already been taken, thus reasonably assuring that the error will not occur again.

§ 40.33 Reporting and review of results.

(a) *Medical review officer shall review confirmed positive results.* (1) An essential part of the drug testing program is the final review of confirmed positive results from the laboratory. A positive test result does not automatically identify an employee/applicant as having used drugs in violation of a DOT agency regulation.

An individual with a detailed knowledge of possible alternate medical explanations is essential to the review of results. This review shall be performed by the Medical Review Officer (MRO) prior to the transmission of the results to employer administrative officials. The MRO review shall include review of the chain of custody to ensure that it is complete and sufficient on its face.

(2) The duties of the MRO with respect to negative results are purely administrative.

(b) *Medical review officer—qualifications and responsibilities.* (1) The MRO shall be a licensed physician with knowledge of substance abuse disorders and may be an employee of a transportation employer or a private physician retained for this purpose.

(2) The MRO shall not be an employee of the laboratory conducting the drug test unless the laboratory establishes a clear separation of functions to prevent any appearance of a conflict of interest, including assuring that the MRO has no responsibility for, and is not supervised by or the supervisor of, any persons who have responsibility for the drug testing or quality control operations of the laboratory.

(3) The role of the MRO is to review and interpret confirmed positive test results obtained through the employer's testing program. In carrying out this responsibility, the MRO shall examine alternate medical explanations for any positive test result. This action may include conducting a medical interview and review of the individual's medical history, or review of any other relevant biomedical factors. The MRO shall review all medical records made available by the tested individual when a confirmed positive test could have resulted from legally prescribed medication. The MRO shall not, however, consider the results or urine samples that are not obtained or processed in accordance with this part.

(c) *Positive test result.* (1) Prior to making a final decision to verify a positive test result for an individual, the MRO shall give the individual an opportunity to discuss the test result with him or her.

(2) The MRO shall contact the individual directly, on a confidential basis, to determine whether the employee wishes to discuss the test result. A staff person under the MRO's supervision may make the initial contact, and a medically licensed or certified staff person may gather information from the employee. Except as provided in paragraph (c)(5) of this section, the MRO shall talk directly with the employee before verifying a test as positive.

(3) If, after making all reasonable efforts and documenting them, the MRO is unable to reach the individual directly, the MRO shall contact a designated management official who shall direct the individual to contact the MRO as soon as possible. If it becomes necessary to reach the individual through the designated management official, the designated management official shall employ procedures that ensure, to the maximum extent practicable, the requirement that the employee contact the MRO is held in confidence.

(4) If, after making all reasonable efforts, the designated management official is unable to contact the employee, the employer may place the employee on temporary medically unqualified status or medical leave.

(5) The MRO may verify a test as positive without having communicated directly with the employee about the test in three circumstances:

(i) The employee expressly declines the opportunity to discuss the test;

(ii) The designated employer representative has successfully made and documented a contact with the employee and instructed the employee to contact the MRO (see paragraphs (c)(3) and (4) of this section), and more than five days have passed since the date the employee was successfully contacted by the designated employer representative; or

(iii) Other circumstances provided for in DOT agency drug testing regulations.

(6) If a test is verified positive under the circumstances specified in paragraph (c)(5)(ii) of this section, the employee may present to the MRO information documenting that serious illness, injury, or other circumstances unavoidably prevented the employee from timely contacting the MRO. The MRO, on the basis of such information, may reopen the verification, allowing the employee to present information concerning a legitimate explanation for the confirmed positive test. If the MRO concludes that there is a legitimate explanation, the MRO declares the test to be negative.

(7) Following verification of a positive test result, the MRO shall, as provided in the employer's policy, refer the case to the employer's employee assistance or rehabilitation program, if applicable, to the management official empowered to recommend or take administrative action (or the official's designated agent), or both.

(d) *Verification for opiates: review for prescription medication.* Before the MRO verifies a confirmed positive result for opiates, he or she shall determine that there is clinical evidence—in addition to the urine test—of unauthorized use of any opium, opiate, or opium derivative (e.g., morphine/codeine). (This requirement does not apply if the employer's GC/MS confirmation testing for opiates confirms the presence of 6-monoacetylmorphine.)

(e) *Reanalysis authorized.* Should any question arise as to the accuracy or validity of a positive test result, only the Medical Review Officer is authorized to order a reanalysis of the original sample and such retests are authorized only at laboratories certified by DHHS. The Medical Review Officer shall authorize a reanalysis of the original sample if requested to do so by the employee within 72 hours of the employee's having received actual notice of the positive test. If the retest is negative, the MRO shall cancel the test.

(f) *Result consistent with legal drug use.* If the MRO determines there is a legitimate medical explanation for the positive test result, the MRO shall report the test result to the employer as negative.

(g) *Result scientifically insufficient.* Additionally, the MRO, based on review of inspection reports, quality control data, multiple samples, and other pertinent results, may determine that the result is scientifically insufficient for further action and declare the test specimen negative. In this situation the MRO may request reanalysis of the original sample before making this decision. (The MRO may request that reanalysis as provided in § 40.33(e) be performed by the same laboratory or, that an aliquot of the original specimen be sent for reanalysis to an alternate laboratory which is certified in accordance with the DHHS Guidelines.) The laboratory shall assist in this review process as requested by the MRO by making available the individual responsible for day-to-day management of the urine drug testing laboratory or other employee who is a forensic toxicologist or who has equivalent forensic experience in urine drug testing, to provide specific consultation as required by the employer. The employer shall include in any required annual report to a DOT agency a summary of any negative findings based on scientific insufficiency but shall not include any personal identifying information in such reports.

(h) *Disclosure of information.* Except as provided in this paragraph, the MRO shall not disclose to any third party medical information provided by the individual to the MRO as a part of the testing verification process.

(1) The MRO may disclose such information to the employer, a DOT agency or other Federal safety agency, or a physician responsible for determining the medical qualification of the employee under an applicable DOT agency regulation, as applicable. only if—

(i) An applicable DOT regulation permits or requires such disclosure;

(ii) In the MRO's reasonable medical judgment, the information could result in the employee being determined to be medically unqualified under an applicable DOT agency rule; or

(iii) In the MRO's reasonable medical judgment, in a situation in which there is no DOT agency rule establishing physical qualification standards applicable to the employee, the information indicates that continued performance by the employee of his or her safety-sensitive function could pose a significant safety risk.

(2) Before obtaining medical information from the employee as part of the verification process, the MRO shall inform the employee that information may be disclosed to third parties as provided in this paragraph and the identity of any parties to whom information may be disclosed.

§ 40.35 Protection of employee records.

Employer contracts with laboratories shall require that the laboratory maintain employee test records in confidence, as provided in DOT agency regulations. The contracts shall provide that the laboratory shall disclose information related to a positive drug test of an individual to the individual, the employer, or the decisionmaker in a lawsuit, grievance, or other proceeding initiated by or on behalf of the individual and arising from a certified positive drug test.

§ 40.37 Individual access to test and laboratory certification results.

Any employee who is the subject of a drug test conducted under this part shall, upon written request, have access to any records relating to his or her drug test and any records relating to the results of any relevant certification, review, or revocation-of-certification proceedings.

§ 40.39 Use of DHHS—certified laboratories.

Employers subject to this part shall use only laboratories certified under the DHHS "Mandatory Guidelines for Federal Workplace Drug Testing Programs," 53 FR 11970, April 11, 1988, and subsequent amendments thereto.

BILLING CODE 4910-62-M

Glossary

Accessioning. A process by which a laboratory receives, identifies, and removes a portion of a specimen for testing while maintaining the specimen's integrity.

Accuracy. The closeness with which results agree with a known true value of the quantity being measured.

Aliquot. A portion of a specimen used for testing.

Amphetamines. A term generally used to include amphetamine and methamphetamine. Other phenethylamines, not all of which are abused, may cross-react with some antibodies used in immunoassay test kits and may be included in this group.

d-Amphetamine. Amphetamine is a specific phenethylamine of known structure that exists in two isomeric forms. The *d*, or dextro form (rotates polarized light to the right), is a potent central nervous system stimulant and is subject to abuse.

l-Amphetamine. The *l*, or levo, isomer of amphetamine (rotates polarized light to the left), is not a potent central nervous system stimulant and is not subject to abuse.

Analyte. The chemical component being measured in an analysis.

Assay. The measurement of the quantity of a chemical component.

Barbiturates. A class of drugs used in medicine as hypnotic agents to promote sleep or sedation. Some are also useful in the control of epilepsy. All are central nervous system depressants and are subject to abuse. Depending on their potency they are classified as Schedule II or Schedule III drugs.

Adapted, in part, from Glossary of terms. In B. S. Finkle, R. V. Blanke, and J. M. Walsh (eds.), *Technical, Scientific, and Procedural Issues of Employee Drug Testing*. Rockville, Md.: National Institute on Drug Abuse; 1990:29–31.

Batch reporting. Urine specimens for drug testing are frequently sent to the laboratory in groups or "batches." Test results are generally reported on all specimens in a batch simultaneously, rather than reporting the negatives first, and then, after a delay while they are confirmed, reporting the positive results. Batch reporting improves confidentiality by helping to avoid identifying those individuals whose tests must be confirmed.

Benzodiazepines. A class of drugs used in medicine as minor tranquilizers frequently prescribed to treat anxiety. They are central nervous system depressants and are subject to abuse.

Benzoylecgonine. A metabolite of cocaine that is readily excreted in the urine; thus, its detection implies cocaine use.

Blind performance testing. When conducted in a blind fashion, the laboratory, and particularly the analyst, is not aware that the specimen being tested has been submitted specifically to monitor laboratory performance. (*See also* **performance testing.**)

Blind QC. Control material that is introduced into a batch of specimens in such a manner that the analyst is unaware that it is not a real specimen. This is done by the laboratory director or by the quality control supervisor in order to make sure that the control material is not given special treatment.

Cannabinoids. The psychoactive substances found in the common hemp plant, or *Cannabis sativa*. Most of the psychological effects are produced by delta-9-tetrahydrocannabinol. In urine drug testing, the prior use of cannabinoids is established by the detection of metabolites of cannabinoids. These are generally inactive but are present in greater quantities than delta-9-tetrahydrocannabinol. The most abundant metabolite is 11-nor-delta-9-tetrahydrocannabinol-9-carboxylic acid, sometimes referred to as 9-carboxy-THC, toward which most immunoassays and confirmation procedures are directed.

Certified copy. A copy of a document (not the original), such as a laboratory report or chain of custody form, which is attested as being a true copy by a responsible official.

Certified laboratory. A laboratory that has met certain minimum performance standards set by an accrediting agency, and has received a certificate to verify this fact.

Certified reference material. A material or substance or drug, one or more of whose property values are certified by a valid procedure, or accompanied by, or traceable to, a certificate or other documentation that is issued by a certifying body.

Chain of custody. Procedures to account for the integrity of each urine specimen, or aliquot thereof, by tracking its handling and storage from point of specimen collection to the final disposition of the specimen. Documentation of this process must include the date and purpose each time a specimen or aliquot is handled or transferred and identification of each individual in the chain of custody.

Chromatography. Any of a variety of techniques used to separate mixtures of drug and their metabolites and other chemicals into individual components based on differences in their relative affinities for two different media: a mobile phase and a stationary phase. In gas chromatography, the mobile phase is an inert gas such as nitrogen or helium and the stationary phase is a high-boiling liquid bound to fine particles packed in a glass column, or bound to the inner surface of a glass capillary column.

Cocaine. An alkaloid, methylbenzoylecgonine, obtained from the leaves of the coca tree (*Erythroxylon* sp.). It is a central nervous system stimulant that produces euphoric excitement; abuse and dependence constitute a major drug problem.

Collection site. A place where individuals present themselves for the purpose of providing a specimen of their urine to be analyzed for the presence of drugs.

Confirmation. The process of using a second analytical procedure to identify the presence of a specific drug or metabolite that is independent of the initial test and uses a different technique and chemical principle from that of the initial test in order to ensure reliability and accuracy.

Consortium. A group or association of employers, operators, or contractors who form together to accomplish drug testing of covered employees.

Creatinine. A substance formed by the breakdown in the body of phosphocreatine and excreted in the urine. The rate of creatinine excretion is a function of body muscle mass and is relatively constant but dependent on the health, sex, and age of the individual.

Cross reactivity. The degree to which an antibody interacts with antigens other than the one used to produce the antibody. This is a property of nearly all naturally derived antibodies.

Cutoff. The defined concentration of analyte in a specimen at or above which the test is called positive and below which it is called negative. This concentration is usually significantly greater than the sensitivity of the assay. (*See also* **limit of detection; sensitivity.**)

Deterrent program. A program, such as a urine drug testing program, that has as its goal deterring individuals from the abuse of drugs.

DHHS. Department of Health and Human Services.

Documentation. A printed or written record retained as support or proof of claims made in reporting test results or in the laboratory certification process.

DOT. Department of Transportation.

Drug metabolite. A modified form or degradation product of a drug produced by a metabolic process.

EAP. *See* **employee assistance program.**

Employee assistance program. A program designed to assist employees with drug abuse or other problems by means of counseling, treatment, or referral to treatment.

False negative. A test result that states that no drug is present when, in fact, a tested drug or metabolite is present in an amount greater than the threshold or cutoff amount.

False positive. A test result that states that a drug or metabolite is present when, in fact, the drug or metabolite is not present or is in an amount less than the threshold or cutoff value.

For-cause testing. *See* **reasonable cause testing.**

Forensic. Suitable for a court of law or public debate or argument.

GC-MS. Abbreviation for the instrumental technique that couples the powerful separation potential of gas chromatography with the specific characterization ability of mass spectrometry.

Immunoassay. The measurement of an antigen-antibody interaction utilizing such procedures as immunofluorescence, radioimmunoassay, enzyme immunoassay, or other nonradioisotopic techniques. In drug testing, the antigen is a drug or metabolite; the antibody is a protein grown in an animal and directed toward a specific drug, metabolite, or group of similar compounds.

Initial testing procedures. The initial test, or screening test, is used to identify those specimens that are negative for the presence of drugs or their metabolites. These specimens need no further examination and need not undergo a more costly confirmation test.

Limit of detection. The minimum amount of an analyte that can be detected with confidence by a testing procedure.

Mass spectrometry. Analysis using an analytical instrument that provides accurate information about the molecular mass and structure of complex molecules. This technique can identify and quantify extremely small amounts of drugs or metabolites by their mass-fragment spectrum.

Medical Review Officer. A licensed physician responsible for receiving laboratory results generated by a drug testing program, and who has knowledge of substance abuse disorders and has appropriate medical training to interpret and evaluate an individual's positive test result together with his or her medical history and any other relevant biomedical information.

d-Methamphetamine. The optical isomer of methamphetamine (desoxyephedrine) that rotates polarized light to the right (dextro) and is the active isomer. It is a central nervous system stimulant and has a strong potential for abuse. Recently it appears to be gaining popularity for illicit use in the form of "ice."

Methaqualone. A hypnotic drug unrelated to the barbiturates category but used as a sedative and sleeping aid. Formerly it was a widely abused drug but seems to be less popular in recent years. It is also known by its trade name Quaalude.

Monitoring. To check constantly on the accuracy and general performance of a laboratory, instrument, or analyst.

NIDA. National Institute on Drug Abuse.

Non-contact positive. When a Medical Review Officer receives a positive test result, extensive efforts are made to contact the employee before reporting the result. When these efforts fail, the result may be reported to the employer as a "non-contact-positive" result.

On-site screening. In those situations in which it is desirable to learn urine drug test results quickly, the preliminary immunoassay screening test may be conducted at the worksite.

Open performance testing. Performance testing that is done with the knowledge of the analyst. (*See also* **performance testing.**)

Open QC program. A quality control program designed by the quality control supervisor or responsible person that is known to the analysts and technologists and is implemented to detect the random and systematic errors that may occur throughout the drug testing process.

Opiate. Term used to designate drugs derived from opium such as morphine and codeine, and the semisynthetic congeners such as heroin. Immunoassay kits for opiates are generally directed to detect morphine but cross-react with other opiates as well.

Passive inhalation. The innocent exposure of nonsmoking subjects to sidestream smoke from active smokers, thereby raising the possibility that a nonuser of marijuana may test positive for metabolites of delta-9-tetrahydrocannabinol.

Performance testing. A program designed to monitor the analytical accuracy and precision of a drug testing laboratory. This is done by periodically submitting challenges of human urine fortified with drugs or metabolites of drugs to the laboratories being monitored. Test results must conform to predetermined limits of accuracy and precision when compared to the test results of reference laboratories or to the mean result of all participating laboratories.

pH. The negative logarithm of the hydrogen ion activity in solution. This is a measure of acidity of a specimen. The lower the number, the more acidic is the specimen. The pH of random urine may range between 4.5 and 8 pH units.

Phencyclidine. One of the most dangerous of the hallucinogenic, illicit drugs, most often referred to as PCP. Psychotic reactions such as extreme anxiety or panic and hypertensive crisis and seizures are common. Many fatalities have occurred through its abuse.

Post-accident test. A drug test administered to an employee when an accident has occurred and the employee performed a function that either contributed to the accident or cannot be completely discounted as a contributing factor in the accident. (The definition of *accident* varies between various federal agency regulations.)

Precision. A measurement of the agreement between repeated measurements. The standard deviation, variance, or coefficient of variation may be used as a measure of precision.

Pre-employment testing. The widespread practice of conducting urine drug testing on applicants for jobs in order to minimize the likelihood of employing a drug abuser.

Presumptive positive. A positive screening or immunoassay test is presumptive and should then be confirmed by a different, more specific test.
PT. *See* **performance testing.**
Qualitative analysis. Relating to a test or measurement that determines the presence or absence of specific drugs or metabolites in the specimen.
Quality assurance. A program in which all procedures and operations of the laboratory are reviewed to ensure optimal operations.
Quality control. A system instituted to maintain the output of a technical operation at a level that has been established as acceptable. It involves the setting of quality standards, continual appraisal of conformance to these standards, and, in the absence of conformance, taking corrective action to establish or maintain the predetermined levels of performance. Both intra- and interlaboratory quality control (QC) are utilized.
Quantitative analysis. The accurate determination of the quality of drug or metabolite present in a specimen.
Random testing. Unannounced, random selection of candidates to be tested.
Real samples. Urine specimens collected from real subjects for testing purposes, in contrast to open or blind PT specimens, control specimens, calibrators, etc.
Reanalysis. A specimen that is taken from storage for a repeat analysis by request of the Medical Review Officer because of a legal challenge.
Reasonable cause testing. Urine drug testing of an individual who is reasonably suspected of drug use based on physical, behavioral, or performance indicators. (Also known as reasonable suspicion testing and for cause testing.)
Revocation of certification. A certified laboratory may have its certification revoked if such a step is necessary to ensure full reliability and accuracy of drug tests and the accurate reporting of drug tests. The factors to consider prior to revocation are variable as are the period and terms of the revocation.
Safety-sensitive positions. Workplace positions that are deemed acutely sensitive to safety considerations such as airline pilots, nuclear reactor operators, train crews.
Screening. *See* **initial testing procedures.**
Sensitivity. The smallest concentration of a drug or metabolite that can be reliably detected by a particular assay method. (*See also* **limit of detection.**)
Specific gravity. The ratio of the density of urine to the density of water at a specified temperature. The specific gravity of random urine specimens ranges between 1.002 and 1.030 at body temperature, depending on fluid intake.
Specificity. The ability of a particular test to identify a drug or metabolite without interference or cross reactions.
Split specimen. The practice of dividing a urine specimen into two portions, one of which may be submitted for analysis and other preserved by freezing for the confirmation analysis or reanalysis.

Target value. The amount of analyte weighed into a specimen during the preparation of a PT specimen that results in an intended concentration. The concentration is confirmed by analysis in a reference laboratory. (*See also* **verified drug concentration.**)

THC. Delta-9-tetrahydrocannabinol, the most active cannabinoid (*See also* **cannabinoids.**)

Threshold. *See* **cutoff.**

Verified drug concentration. The confirmation that a target value has been achieved in preparing PT specimens, calibrators, or blind controls. This is generally done by careful analyses by reference laboratories.

Index

Absenteeism, relation to drug use, 8-11, 33
Accessioning, 83
Alcohol abuse, workplace programs, 141-142
Alcohol testing, 16, 42, 78-79, 93
  MRO role, 128
Americans with Disabilities Act, 19-20
Amphetamine, laboratory testing, 88-89
Armed Forces Institute of Pathology, laboratory monitoring program, 27

Barbiturates
  laboratory testing, 92
  use, 197
Benzodiazepines
  laboratory testing, 92
  use, 189, 198-200

Calibrators, quality control, 166-167, 173
Chain of custody forms, 56, 58-59, 98-99, 186
  chain of custody block, 56, 58, 105, 108
  distribution of, 58-59, 98-101, 195-196
  DOT, 58-59, 99-100
    MRO block, 124, 126, 199-200
  errors, 76-78, 105-107, 195-196
  internal chain of custody form, 84, 98
  review of, 100-101, 105, 198-199
Challenging test results, 43, 200-201
Coast Guard regulations, 4-5
  results reporting, 124
Cocaine
  laboratory testing, 87-88
  passive inhalation, 122
  skin absorption, 197
  use, 5-6, 31, 114, 196-197
Codeine, laboratory testing, 89-90, 196-197
Collection procedures, 53-79
  alcohol testing
    blood, 79
    breath, 78-79
  chain of custody errors, 76-78, 105-107, 195-196
  checklists, 61-66
  forensic standards, 53-54
  notifying donors, 61
  packaging standards, 75-76
  parallel (multiple) collections, 68
  problem solving, 68-75
    donor not bringing photo identification, 68-69
    donor not cooperative, 74
    donor not reporting to collection site, 68
    donor requiring medical attention, 74
    less than 60 ml of urine, 72-74
    medical examiner's certificate and drug testing, 74-75
    monitored collections, 69

239

Collection procedures (*continued*)
  witnessed collections, 69–72
  recordkeeping, 56, 78
  specimen security, 76
  split specimens, 67–68
  transport to laboratory, 76, 195–196
Collection services, costs, 47
Collection site preparation, 54–61
  chain of custody forms, 56, 58–59
  consent and release forms, 59–60
  drug test collection record book, 56–57
  medication lists, 59, 61
  selection of, 54–55
  supplies, 55–56
  water sources, 55
College of American Pathologists/American Association of Clinical Chemistry. *See* Laboratory certification
Confidentiality, 38
  drug testing, 45, 102, 138–139, 194–196, 200–201
  employee assistance programs, 143, 160
Confirmation testing, 43, 85–86
Consent and release forms, 59–60
  witnessed collections, 69, 72
Cost-benefit analyses
  drug testing programs, 32–33, 35, 46–49
  employee assistance programs, 151–153, 159
Costs
  drug testing programs, 46–47
  external blind performance testing, 176–177
  laboratory certification, 183, 185
Cross-reactivity, laboratory testing, 89

Department of Defense
  defense contractors, drug-free work force, 3–4, 29
  drug testing regulations, 29
Department of Energy regulations, drug testing, 30
Department of Transportation Anti-Drug Information Center, 21
Department of Transportation drug testing procedures, 29, 219–229
  chain of custody forms, 58–59, 99–100, 105
  external blind performance testing, 175
  medical fitness-for-duty information disclosure, 112–113
  recommended test invalidation criteria, 105–106
  results reporting, 123–126
  routing of laboratory-negative results, 102–104
DHHS Mandatory Guidelines for Federal Workplace Drug Testing Programs. *See* NIDA Guidelines
Donor
  certification signature, omitted, affidavit for, 105–107
  collection checklist, 62
  confirming identity
    at collection site, 68–69
    during review of results, 112
  contacting to review positive results, 109–112
  informing of test results, 113
Drug abuse in the workplace, 1–20
  economic costs, 10–12, 31–33
  employer responses to, 13–15, 34–35

federal programs and policies prior to 1990, 1–5
information sources, 20–21
on-duty versus off-duty use, 38
performance and productivity effects, 8–13, 32–34
prevalence, 5–8, 31–34
treatment, 48, 145–146, 148
Drug-Free Workplace Act of 1988, 2–3
educational program, 145
employee assistance programs, 142
Drug-free workplace program, components, 10, 12–13, 36–37
Drug testing, 23–49
case studies, 193–202
collection procedures, 195–196
MRO function, 197–200
pharmacologic issues, 196–197
performance testing, 201–202
programmatic issues, 193–195
risk management, 200–201
challenging results, 43
classes of drugs, 42, 189
confidentiality and privacy protections, 38, 45, 200–201
consequences for positive results, 44–45
costs, 46–47
court rulings, 42
employee morale impact, 48–49
ethical guidelines, 205–206
evaluating need for, 31–36
extent of use among work-force, 31–32
impact of use among work-force, 32–34
program benefits, 35–36
purpose of testing, 35
history, 23–30

DOD regulations, 29
DOE regulations, 30
DOT Procedures, 29
military, 25–27
NIDA Guidelines, 27–28
1980s, 24
1960s and 1970s, 23–24
NRC regulations, 29–30
private-sector, 24–25
state and municipal laws, 30
legal framework, 134–135
military, 25–27
on-duty versus off-duty use, 38
on-site, 188
versus off-site, 43
periodic, 40–41
policy, 36–46
communication to employees, 39
elements, 36–37
key issues, 37–38
post-accident/post-incident, 40
postrehabilitation/extended absence, 40, 151, 155, 158
preemployment/preappointment, 39
prevalence, 13–14, 34–35
random, 41–42
reasonable cause, 39–40
records, 78, 138
employer recordkeeping, 45–46
retention, 126, 128
safe harbor, 2, 43–44
what to test for, 42–43
who to test, 37–39
Drug Testing Quality Act (H.R. 33), 18–19, 30, 189

Educational programs in the workplace, 10, 12, 145, 160

Employee Assistance Professionals Association, 21
Employee assistance programs, 10, 12–13, 47–48, 141–161
confidentiality, 143, 160
cost-benefit analyses, 151–153, 159
description, 143–147
drug abuse services, 145–146, 148
history, 141–142
monitoring recovery, 151–158
problems addressed by, 143–144
program accessibility, 159–160
program services, 160
quality assurance, 160
referral guidelines, 148–150
selection, 159–161
staff qualifications, 159
Ephedrine, causing false positive test results, for methamphetamine, 89, 201–202
Ethical guidelines, drug testing, 205–206
Executive Order 12564, 1–2, 27–28
drug-free workplace program components, 10, 12–13
implementation guidance, 36–37
Extended absence drug testing, 40

Federal Aviation Administration, 4–5
results reporting, 123–125
Federal drug abuse programs and policies, 1–5, 15–20, 25–30
Federal Highway Administration, 4–5
conditions for acceptable medication use, 121
medical examiner's certificate, 74–75, 193–194
Federal Railroad Administration, 4–5
conditions for acceptable medication use, 121
results reporting, 123
Fitness-for-duty information, disclosure, 112–113
Forensic laboratory. *See* Laboratory selection

GC/MS analysis, 86
laboratory inspections, 187

Hair analysis, 93–94
Hatch-Boren bill. *See* Quality Assurance in the Private Sector Drug Testing Act
Health, employee, drug testing impact, 48
House Resolution 33. *See* Drug Testing Quality Act

Identification, donor not bringing, 68–69
Immunoassay analysis, 84–85
laboratory inspections, 186
Internal chain of custody document, 84
International Olympic Committee, drug testing, 189

Laboratory analysis, 83–87
confirmation testing, 85–86
costs, 47
GC/MS analysis, 86
immunoassay analysis, 84–85
Laboratory certification, 28, 81–82, 164, 171–173, 183–184
Laboratory inspections, 166, 182–188
checklist and critical questions, 183–184
equipment and equipment maintenance, 187

frequency, 183
GC/MS, 187
immunoassays, 186
inspection team, 183
personnel, 187-188
protocols and reports, 183-184
quality control, 187
reagents, 187
records, 188
resources, 184-186
specimen handling, 186
standard operating procedures, 186
Laboratory performance, monitoring, 163-190. *See also* Performance testing
accuracy concerns, 163-164
case studies, 201-202
future, 188-190
laboratory inspections, 166, 182-188
quality assurance, 165, 169-171
quality control, 165-169
Laboratory selection, 81-83
Laboratory testing
alcohol, 93
amphetamine and methamphetamine, 88-89, 201-202
barbiturates, 92
benzodiazepines, 92, 189
cocaine, 87-88
codeine, 89-90, 196-197
hair analysis, 93-94
marijuana, 87
morphine, 89-91
phencyclidine, 90-91
results, reporting to MRO, 101-104
sports medicine, 189

Managed mental health-care providers, 142
Mandatory Guidelines for Federal Workplace Drug Testing Programs. *See* NIDA Guidelines
Marijuana
laboratory testing, 87
passive inhalation, 122
use, 5-6, 31
Medical benefits usage and drug use, 10, 11
Medical examiner's certificate, 74-75, 193-194
Medical Review Officer (MRO), 97-129
alcohol testing, 128
case studies, 197-200
chain of custody form, receipt and review, 98-101
client setup checklist, 137
communication with donor, 139
confidentiality, 138-139
continuing education, 97-98, 139-140
contracts, 136-138
costs for services, 47
definition, 97
insurance coverage, 140, 200-201
invalidating tests with significant errors, 104-108
laboratory result receipt, 101-104
MRO negatives, 102
professional demeanor, 139
punchlist, 109-111
qualifications, 97-98
recordkeeping, 126, 128, 138
rehabilitation and return-to-work determinations, 128-129
reporting results to employer, 123-127

Medical Review Officer (*continued*)
  reviewing laboratory-positive results, 108–123
    contacting donor, 109–112
    determination of methamphetamine isomeric form, 119
    donor's identity confirmation, 112
    fitness-for-duty information disclosure, 112–113
    inform donor of test results, 113
    legitimate drug use, 113–117, 120–123
    opiate-positive results, interpreting, 117–118, 196–198, 200
    potential sources of drugs, 113–116
    quantitative results, 119
    reanalysis, 119–120
    split specimen analyses, 87, 119–120, 198–199
    spousal use, 122, 197–199
    verify drug sources, 114, 116–117
Methamphetamine
  determination of isomeric form, 119
  laboratory testing, 88–89
  skin absorption, 197
Military drug testing, 25–27
Monitored collections, 69
6-Monoacetyl morphine (6-MAM), 90, 117–118
Morphine, laboratory testing, 89–91
MRO. *See* Medical Review Officer
Municipal laws, drug testing, 30

National Association of State Alcohol and Drug Abuse Directors, 21

National Clearinghouse for Alcohol and Drug Information, 20
National drug control strategies, 15
NIDA Guidelines, 2, 27–28, 207–217
  batch reporting of test results, 101
  certification program, laboratory, 81–82, 164
    performance testing, 171–173
  collection procedures, 53
  confirmation testing, 86
  drug testing cutoff concentrations, 85
  external blind performance testing, 174–175
  MRO definition, 97
  open performance testing requirements, 171
  quality control sample requirements, 167
  retention of specimens testing positive, 86
NIDA Toll-Free Drug Abuse Information and Treatment Hot Line, 20–21
NIDA Toll-Free Workplace Help Line, 20
NRC regulations, 4, 29–30
  batch reporting of test results, 101
  conditions for acceptable medication use, 121
  external blind performance testing, 175

Occupational alcoholism programs, 141–142
Omnibus Transportation Employee Testing Act of 1991, 16–17, 189

On-site drug testing, 43, 188
Open controls, quality control, 167–168
Opiate abuse, clinical evidence, 117–118

Packaging standards, specimens, 75–76
Parallel collections, procedures, 68
PCP
  laboratory testing, 90–91
  passive inhalation, 122
  use, 114
Performance, drug use and, 8–13, 32–34
Performance testing
  case studies, 201–202
  drug concentration validation, 176
  external blind, 165–166, 174–182
    collection site errors, 180
    costs, 176–177
    definition of unsatisfactory performance, 182
    encoding, 177
    laboratory errors, 181
    reasons for sample discrepancies, 178–181
    requirements, 174–175
    result recipient errors, 181
    specimen preparation, 175–176
    tracking results, 178–179
    vendor sample errors, 180
  internal blind, 168–169
  open, 167–168, 171–174
Periodic drug testing, 40–41
Phencyclidine. See PCP
Poppy seeds, ingestion, 90, 117
Positive test results
  consequences, 44–45
  interpreting. See Medical Review Officer (MRO), reviewing laboratory-positive results
  retention of specimens, 86
Post-accident/post-incident drug testing, 40
Postrehabilitation drug testing, 40, 151, 155, 158
Preemployment/preappointment drug testing, 39
Prescription drugs
  drug testing, 42
  medication lists, 59, 61
  positive test results, 113–116
Privacy. See Confidentiality
Private-sector testing, 24–25
Productivity, relation to drug use, 8–13
Public Law 100-71, 2

Quality assurance
  definition, 169
  equipment maintenance, 170
  laboratory procedures, 165, 169–171
Quality Assurance in the Private Sector Drug Testing Act (S. 1903), 17–18, 98, 189
Quality control
  calibrators, 166–167
  internal blind control samples, 168–169
  internal laboratory procedures, 165–169
  laboratory inspections, 187
  open controls, 167–168

Random drug testing, 41–42
Reagents, laboratory inspections, 187
Reanalysis of specimens, 86, 119–120

Reasonable cause drug testing, 39–40
Recordkeeping, 138
  employer, drug testing, 45–46
  laboratory inspections, 188
Recovery, monitoring, employee assistance programs, 151–158
Recovery contract, 156–157
Recovery plan agreement, 155
Rehabilitation, MRO role, 128–129
Research and Special Programs Administration, 4–5
Return-to-work conference, 151
  checklist, 154
  definition, 152
  guidelines, 153
Return-to-work determinations, MRO role, 128–129
Risk management strategies, 133–140
  assisting clients with program development, 135–140
  case studies, 200–201
  communication, 139
  confidentiality, 138–139
  continuing education, 139–140
  contract use, 136–138
  documentation, 138
  insurance coverage, 140
  standardization, 138

Safe harbor, 2, 43–44
Safety issues, drug testing's role, 37, 48
Security issues, drug testing's role, 37
Senate Bill 1903. *See* Quality Assurance in the Private Sector Drug Testing Act

Specimen
  adulterated or diluted, 105–106, 108
  handling, laboratory inspections, 186
  preparation, performance testing, 175–176
  reanalysis, 119–120
  security, 76
  transport to laboratory, 76, 195–196
Split specimens
  analyses, 87, 119–120, 198–199
  collection procedures, 67–68
  positive test results, 87
State laws, drug testing, 30
Substance abuse, treatment of, 48, 145–148
Supplies, collection site, 54–55
Survey of Employer Anti-Drug Programs, 13–14

Transportation Workplace Drug Testing Programs. *See* Department of Transportation drug testing procedures

Urban Mass Transit Administration, 4–5
  drug testing, 29

Vicks Inhaler®, positive methamphetamine result, 88, 119

Water loading, 108
Westgard quality control rules, 167–168
Witnessed collections, 45, 69–72
  authorization form, 69, 71
  circumstances for, 69–70
  consent form, 69, 72

Made in the USA
Monee, IL
27 November 2020